BLUE COLLAR INFANTRYMEN

REFLECTIONS ON 35 YEARS OF SOLDIERING

BLUE COLLAR INFANTRYMEN

REFLECTIONS ON 35 YEARS OF SOLDIERING

BRIAN REED, BRIGADIER GENERAL
U.S. ARMY, RETIRED

DEEDS PUBLISHING | ATHENS

Published by Deeds Publishing in Athens, GA
www.deedspublishing.com

Printed in The United States of America

Cover and interior design by Deeds Publishing

ISBN 978-1-961505-53-7

Books are available in quantity for promotional or premium use.
For information, email info@deedspublishing.com.

First Edition, 2026

10 9 8 7 6 5 4 3 2 1

For
Kirsten, Stephanie, Kaelin, and Brandon

Dedicated to the Blue-Collar Infantryman
Strength and Honor

DISCLAIMER

This narrative is drawn from memory. The views expressed are strictly my own and do not reflect the position of the U.S. Army or the Department of Defense. I own the stories, and I own the mistakes.

CONTENTS

PREFACE

I was in the Dining Facility (DFAC) at Camp Vance in Bagram, Afghanistan in June 2019. I was a senior Colonel in the United States Army at the time... just over 30 years. I was stationed at the United States Military Academy, so I definitely didn't need to be here. The truth was, I wasn't sure why I was there. My wife of 29+ years at the time told me on the way out the door when I headed to the airport, "I hope you find what you are looking for." I guess I may have been in some sort of mid-life crisis. Who knows... I sure as hell didn't know.

All I knew was that a good friend called me in April of the same year and asked me if I wanted to head to Afghanistan with him. He was taking over as the Commanding General for the Special Operations Joint Task Force-Afghanistan/NATO Special Operations Component Command-Afghanistan (SOJTF-A/NSOCC-A). There was more to the conversation than that, but I pretty much said yes (without really consulting my wife... who is a saint... and has put up with me all these years) and off I went for several months.

That's how I found myself sitting in the DFAC one morning waiting to meet an officer for breakfast. Major (MAJ) Matt B. had worked for me in a previous assignment and now he was the Battalion Executive Officer of an Infantry Battalion in the 10th Mountain Division. At the time, his unit was responsible for providing security at Camp Vance and Bagram Airfield, as well as a Quick Reaction Force in case the "shit hit the fan." Like me, Matt was a career Infantryman, albeit with a lot less years in.

Over mess hall coffee, eggs, hash browns, and bacon we caught up, talked about old times, and the future. As an aside, in my humble opinion, there is nothing better than breakfast in an Army Mess Hall (or more correctly called the DFAC) — the breakfast of champions! At one point, Matt looked at my sidearm and asked me why I was using the standard issue Beretta 9mm... the same pistol I had been issued at various times throughout my career. While never the primary weapon of an Infantryman, its purpose is as a personal security weapon. Something to pull out in a close gunfight or pistol-whip the shit out of a bad guy... neither of which I had ever done in my Army career to that point!

He asked because I was attached to a special operations unit, and the personal weapon of choice is never... or very, very rarely... the 9mm. There are several reasons for that, but let's just say I had the option to carry something lighter, sleeker, and definitely more high speed. The problem, as I told the 1SG who issued me the weapon, is I would probably accidentally shoot myself with the pistol he wanted me to use. "Give me the old school 9mm," I told him! It's safer for everyone!

To that point in our careers, neither Matt nor I had spent a minute inside a special operations formation. While we both worked with special operations units multiple times during the Global War on Terror — and even though I was now attached to such a unit for the time I was in Afghanistan — we were both... in Matt's words: "Just 9mm-toting, blue-collar Infantrymen." So, a shout out to Matt B. for the title of this book, but more importantly, for helping me to figure out who I was as an Army professional.

See, up until that moment, I always had this thought in the back of my mind that I needed to be in a Special Operations formation — or a Special Operator — to truly be at the tip of the spear in my profession... especially one as an Infantryman.

As a Lieutenant, I was convinced I needed to go into the Special Forces. I did not. As a senior Major, I assessed for the Ranger Regiment... not selected. I also assessed for the Army's elite Tier 1 Special Mission Unit... not selected. Hell, I recycled two phases of Ranger School. It took me 30 years, but I finally figured out that being an Infantryman — and I think a pretty good one at that — wasn't such a bad gig.

So... this is my story, but more importantly the stories of those Infantrymen with whom I served. While we never may have been selected, or even assessed, as special operators, we were special operators. We were hard as woodpecker lips. We were experts of our craft. We were Professional Soldiers. This is the story of the Blue-Collar Infantryman.

1. THE BEGINNING

UNITED STATES MILITARY ACADEMY

When asked why I decided to attend West Point, there's a bit of a backstory with a boating accident while in high school (no one was killed or injured), a rescue by the Coast Guard, a desire to attend the Coast Guard Academy, and then I ended up at the United States Military Academy.

As a kid, my dad took me to a few Army-Navy games in the 1970s. I grew up in the Philly area, and at the time, the game was played in Philadelphia every year... so I suppose that was my first taste of the Service Academy experience. While my dad served in the Army during the Vietnam War period, and I actually spent some time living at Ft. Carson and Ft. Irwin in the late 60s and early 70s, I really have no memory of that.

At the end of the day, I suppose I joined the Army and went to West Point because I wanted to challenge myself and I wasn't really sure what I wanted to do with my life.

Like most new cadets, I was seriously reconsidering my life's choices when I reported for R-Day (or Reception Day) on 1 July 1985. I really had no clue about what I was getting myself into. There was no internet or social media, so you were pretty much flying blind back in the day. Any intelligence or advance warning I had on what West Point was all about was gathered from talking to others and a candidate visit. In short, I had no idea what awaited me.

I walked into a buzzsaw on Day 1. I was immediately assigned to 4th Platoon, Delta Company — Company D4, the Dukes! From R-Day to Graduation, we stayed together. While there are certainly pros and cons to all that, I only see the pros and the shared experiences that bind us together to this day.

I initially underachieved at the Academy. Academically, I struggled my first three semesters, and while I never failed a course, I was perilously close. I was a good student in high school and had to work hard, but it came fairly easy. The volume, pace, and degree of academic difficulty at West Point kicked me in the ass.

I also thought the regulations, rules, and policies were a guide or suggestion, and not necessarily something for me to follow. I was never a crazy, out of control kid in high school, but I did like to have a good time. This all caught up with me on several occasions while at West Point, and I was rewarded with reflective time on the area (i.e., walking hours or area tours), room restriction, loss of privileges, etc. None of the offenses were catastrophic and were normal college shenanigans, but at West Point it was like you sold the state secrets.

Needless to say, life is about choices and choices have consequences. I walked the area for some of my poor choices!

The turnaround for me was the first semester of my sophomore, or yearling, year. My decision making and stupidity ("we'll never get caught") resulted in no Thanksgiving leave and on the boomerang — or prison bus — for the Army-Navy game, the result of something like 90 days of restriction and loss of privileges. The good news was that I now had dedicated time to get my head out of my ass. Things turned around for me academically.

Beginning with the second semester of my yearling year, I made the Dean's List every semester until graduation. Unfortunately, my GPA was so poor the first three semesters, I graduated with a 2.79 GPA, which meant when it came to eventually going to graduate school (and, oh by the way, earning my Ph.D. in Sociology) I was initially on a probationary acceptance.

The good news is that I had matured, was more disciplined, and I loved what I was studying in graduate school, so it was no issue. The point is that I could have made things easier on myself at West Point if I would have gotten my shit straight at the start.

While my decision making when it came to cadet jackassery got better over time, it was never perfect. Fortunately, future shenanigans stayed below the radar, and I never found myself in front of the man again for violating rules or policy. It's funny because much, much later in my career, when I became the Brigade Tactical Officer at West Point, my classmates asked me how I kept a straight face when I gave a cadet hours for some "grievous" violation of the Cadet Disciplinary Regulations.

As I told many cadets when imposing justice at a Brigade Board: "Here's your 100 hours. Don't worry. Someday you can grow up to be a Colonel."

When it came time for branching my senior or firstie year, there was really no doubt that I was going to branch Infantry. I thought a bit about Armor because of my mentor, then-CPT Bob Cone, but my heart was with the Infantry. What sealed the deal for me was the military training and events like Sandhurst, Camp Buckner, Cadet Troop Leadership Training (CTLT) in Germany, Airborne School, Jungle School in Panama... generally all the summer and Army-specific training. I loved the toughness of it; the shared suck; the camaraderie and cohesion of doing something hard with teammates. And I wanted to be at the tip of the spear.

When I threw my hat in the air on a rainy — very rainy day — at Michie Stadium on a random Saturday at the end of May in 1989, I was glad to be done. It was almost unreal, like an out of body experience. But I had no intention of making the Army a career. I was engaged to be married, excited about my personal and professional future, and looking forward to my first assignment in Germany. There was no doubt in my mind that I was a "5 and fly" guy — do my 5 years active-duty service obligation and then go make millions.

That was the plan. Something happened, though. Along the way I found out that I loved Soldiers and I loved being a Soldier... and I didn't leave for 35 more years.

THE INFANTRY OFFICER BASIC COURSE AND RANGER SCHOOL

After a few weeks of post-graduation leave, I began the Infantry Officer Basic Course (IOBC) at Ft. Benning, Georgia in mid-June 1989. Most graduating cadets take as much leave as possible; in some cases, close to 60 days. I decided on going with the earliest IOBC date. Looking back on it, this was partly because I was ready to get started; partly because my roommate and good buddy was going to that course; and partly because I completely over thought it.

Specifically, I figured I'd get to Ranger School quickly, be done, get married (I was already engaged), get to my unit early, and miss the worst part of the winter while in Ranger School. That logic train turned out to be a big fail, especially as it came to Ranger School.

To be honest, IOBC was unremarkable — which is why I'm dedicating five whole sentences to it, including this one. I don't recall much about it, other than that I graduated. I didn't exactly set the world on fire, but I earned my blue cord, and I was a bona fide Infantry officer. Next stop... Ranger School. But first, I was getting married.

I'm not a marriage counselor, but I don't recommend getting married six days before Ranger School. That said, here we are over 35 years later, and going strong! I graduated IOBC on a Friday morning, flew to Philadelphia that afternoon, barely arriving in time for my rehearsal dinner that night; got married the next morning on September 30th; and flew to the Bahamas on Sunday, returning to Ft. Benning, GA. on Wednesday evening of that same week. It was a short honeymoon, and I was back just in time to get my head shaved and complete final preparations prior to reporting to Ranger School with Class 1-90 on a Friday morning in early October 1989.

So, to recap... I graduated from the IOBC on Friday, September 29, 1989, and I began Ranger School a week later, Friday, October 6, 1989, with Ranger Class 1-90. Fast forward, I graduated with Ranger Class 3-90 on March 13, 1990. That's right... two recycles and 159 days later. As it turned out, recycling two phases of Ranger School was arguably the most formative experience of my Army career, both as a leader and an Infantry officer.

When I went to Ranger School, there were four phases — the Darby Phase at Ft. Benning, GA; the Mountain Phase at Dahlonega, GA; the Florida Phase at Eglin Air Force Base (AFB); and the Desert Phase at Dugway, Utah. I also went to Ranger School in the winter, which added a whole other dimension to the experience. On my part, there was never a serious consideration that I would quit Ranger School, although there were several times I felt very, very sorry for myself.

The Darby Phase was generally uneventful for me. I did not have any issues with patrols, physical fitness, the foot marches, the obstacle courses, etc. It was pretty much a smoke-fest where we did a lot of physical training (PT), learned the basics of patrolling and small unit tactics, and we Ranger students were graded on squad level patrols. I was young, physically fit, full of motivation, and had a high pain tolerance. It was here that I learned "pain is weakness leaving the body" was a real thing!

The Mountain Phase started off well. The first part, mountaineering, was tough, but I had no issues — knots, basic and advanced mountaineering, rappelling. The move into the second part of the Mountains, platoon operations, was when I ran into problems. Unlike Darby where the patrols are in squad-size, roughly 9-10 Rangers — in the Mountains we patrolled as a platoon of 30-40 Rangers. The scope of responsibility for a Ranger in a leadership position is magnified significantly.

I did not pass a patrol as a Platoon Leader (PL) or a Platoon Sergeant (PSG), which was a requirement to advance. I had two chances at this, and both were as a PL. It was really no surprise that I did not pass either one of the patrols.

The first opportunity was during the planning and initial movement of the patrol. We had a problem in land navigation from the Patrol Base to the Objective Rally Point (ORP), and as a result we were late and behind schedule. We got lost. As the PL, this was my responsibility... no go. My second opportunity occurred when the entire chain of command was fired midway through an ambush mission.

My roster number was called (something every Ranger dreaded during a mission gone bad) and I was now the new PL. I could never establish personnel accountability, and we had to stop every few minutes of our movement to the ambush site to make sure we had all the Rangers. In this "movement to daylight", everyone was smoked — exhausted — and whenever we stopped, guys were falling asleep. It was a mess. The ambush eventually happened, way off schedule, and it was the end of a nightmare. I earned my second no go.

During my recycle and extended stay (about 10-days) in Dahlonega, I performed "kitchen police" or "KP" duties cleaning pots and pans, scrubbing tables, and the like; learned how to operate a buffer — definitely a perishable skill; and did a lot of police call — picking up of trash and road kill — as I waited with the other recycles for the next class. It was a humbling experience and important in my development as an officer.

To the latter point, I'm not kidding. It gave me an appreciation for what it is like to be at the bottom of the hierarchical food chain and the dirty, nasty jobs that someone has to do. This was probably the first step in building empathy into my leadership style.

After the next class showed up, the recycles integrated into their new companies with Ranger Class 2-90. Other than spending Thanksgiving Day on Mt. Yonah, it was a good phase. I was good-to-go the second time through. I got one PL patrol. That was all I needed. Two things made the difference: (1) I was much more deliberate in attention to detail when it came to navigation. I did not rely solely on the point man and pace man. I stayed dialed-in so I could course correct if necessary; and (2) I was more hands-on... literally.

When a Ranger was not focused, falling asleep, or simply out of it, I would grab him by his LBE (Load Bearing Equipment Vest) or his shoulders and make him look me right in the eyes as I spoke to him with purpose and directness. Both lessons served me well throughout the rest of Ranger School.

The good news about the Florida Phase was that I only needed to do it once. While very tough, things went well in Florida. Because of my Mountain recycle, I was now in Florida in December, which made for a very cold time during waterborne operations. I pretty much froze my ass off and it was the first place I suffered a cold weather injury — go figure, in Florida. After a few hours in the warming tent, I was declared "good to go" and off I went. There was no cancellation or re-scheduling of any of the missions... we went for it and executed to standard despite the cold water and air temperatures.

We spent a lot of time in the water in Florida. Waterborne operations ruled the day. I learned some good lessons about mental toughness in Florida that I have never forgotten. I dug deep into places within me that I didn't know existed. When we finished with Florida, our class was bussed back to Ft. Benning, GA.

Because my Mountain Phase recycle rolled me from Class 1-90 to 2-90, I now had the Christmas Exodus and a break until we began the Desert Phase in Dugway (it may have been two weeks, but I am not for certain). I returned to Ft. Benning in early January 1990 to begin the last phase of Ranger School. I arrived completely out of shape and paid for it with the rest of my class that first day.

All I did over the break was overeat — I had the perfect "food baby." Too many trips to the Shoney's buffet line will do that! After a thorough shake down and a session of push-ups, flutter-kicks, and burpees that was completely miserable, we packed our gear, boarded a C-141, performed an inflight rig, and jumped into the cold and snow at Dugway, Utah. My memory of that airborne operation is that I jumped the M60 Machine Gun, strapped to my leg, and neither the M60 or my rucksack released, so I rode them to the ground. I'm lucky I didn't break a leg!

The Desert Phase was essentially two-parts: force-on-force and live fire. Generally, it was widely considered among Ranger students that if you got to the Desert, you were going to graduate. I had two patrols in a leadership position to get the one "go" I needed. The first was as a PSG during the force-on-force phase. While setting up in our planning patrol base, the platoon got "blown out of there" by enemy artillery. It was a hasty and chaotic relocation that resulted in about a one kilometer movement at dusk.

When we conducted a sensitive items' check at our new location, we were down one pair of Night Vision Goggles (NVGs). In the chaos of our movement, one Ranger left his NVGs — a sensitive item — at the initial patrol base. Even though we found the sensitive item, the Squad Leader (SL) and I each received a no go for the patrol.

This was a bitter pill to swallow. Even though I did not lose the NVGs, I was still responsible since I was the PSG. From my earliest days as a cadet at the U.S. Military Academy, I was taught that the leader is responsible for everything that the unit does or fails to do. Regardless, it did not make me feel any better. I had no recourse. I now needed to get a go on my last patrol. My final opportunity was literally on the last day of the Desert Phase. It was a squad leader patrol during the live fire evolution.

To this day, I do not know why I received a no go. Perhaps I was too nonchalant or overconfident, and my leadership reflected that. I knew I needed a go, so I had a sense of purpose and urgency, but I also knew that so much of the live fire phase was scripted for safety considerations. Maybe this made me take things for granted. As an example, the movements were not particularly onerous and where you placed your support by fire or assault force did not require much creativity. These were generally regulated locations due to range safety requirements.

Regardless, I failed this patrol. This stung... and it hurt badly. I tried to rationalize in my own mind what a Lack of Motivation (LOM) would look like if I just signed it and quit — and how would I explain that I quit Ranger School to family, friends, and my new unit. How would I explain it to myself? Or to my kids someday?

I did not think I could endure another recycle. This was the lowest I ever felt in my young career. At the end of the day, I did not quit. I just could not envision a scenario where I voluntarily said "enough...I quit". Even though I felt I was the victim of a bad Ranger Instructor (RI), I knew I had to keep going. Not once during this process did I ever consider that the reason I failed my patrols was due to my own actions... or inactions. It would not be until later in my career when I would accept responsibility. Regardless, I stayed and sucked up the recycle.

As a handful of my classmates and I waited to see the 7th Ranger Training Battalion leadership for our official notification of a recycle, the rest of our class boarded aircraft for the trip back to Ft. Benning and the graduation jump into Fryer Drop Zone (DZ). Due to the scheduling around the Best Ranger Competition, the follow-on class was delayed, and I had to wait about five weeks at Dugway for the next class to arrive. Again, I spent a lot of time doing police calls, operating a buffer, and picking up tumbleweeds... more development. Once Class 3-90 rolled in, I linked up with the class, took nothing for granted, and finally earned my tab on March 13, 1990.

While I was incredibly proud to have earned the right to be called an Army Ranger, I was also embarrassed that I had recycled twice. For many years, I tried to hide this by not talking about the details of my actual Ranger School experience. It was not until battalion command that I really let go of my embarrassment of being a two-time recycle. I realized that my story was one of grit and resilience, and mental toughness. As I sent Soldiers off to Ranger School as a commander, I could truly tell them that they could do this if they did not quit. I tried to send as many young Soldiers, Non-Commissioned Officers (NCOs), and Officers to Ranger School as possible. I believed that we were a better unit with more Rangers in our formation. I tried to use my story as one of inspiration and motivation.

Fun fact about that RI who I thought screwed me over in the Desert Phase. Several years later, I was stationed at Schofield Barracks in Hawaii. My neighbor was a Chief Warrant Officer 2 (CW2) and a Blackhawk pilot in the Aviation Brigade. Reggie and I became good friends. Our kids played together. One day we were outside of our houses on post — watching the kids play while we had a couple beers... maybe not the best combination. Anyway, we went into his house to grab another beer, and I was looking at the pictures on his wall in his living room.

There was a battalion picture of all the RIs in 7th RTB taken in March 1990. I asked Reggie about the picture and when he arrived as a RI. As it turns out, he was a brand-new RI, and a junior one at that as a Sergeant (most RIs were more senior NCOs). The RI that gave me a no go on my patrol was also a Sergeant and had the same last name as Reggie. What the fuck! That's right... then-Sergeant Reggie F. was my RI, and he was the one who failed me. His only response to me was that I didn't meet the standard and I deserved it. Lesson learned! Responsibility taken. And a reminder that it is a small Army.

2. FIRST ASSIGNMENT: GERMANY

MEETING MY PSG — SFC FRANK G... AND SSG E'S 3 PRINCIPLES

After finally getting out of Ft. Benning, I reported to my first duty station in June 1990 — 1st Battalion, 16th Infantry Regiment, 1st Infantry Division (Forward). The Iron Rangers. I had finally arrived. We were "forward", because the unit was forward based in Boblingen, Germany at Panzer Kaserne... just outside of Stuttgart. There was one brigade "forward" in Germany. The rest of the Division was back at Ft. Riley, Kansas along with the Division Headquarters. It was here in 1-16 IN where I really began my Army career. It will always be my first!

I'm not going to lie... I was feeling pretty good about myself. Despite my setbacks at Ranger School, I was an Airborne Ranger. I felt like I was ready to go. All of that false bravado quickly melted away when I walked into the headquarters of Bravo Company, 1-16 IN, and the real Army! It was here where I met my company commander, Captain (CPT) Doug B., who turned out to be a great commander and was a friend for many years. More importantly, however — and taking nothing away from Doug — I met the man who would shape me, my first Platoon Sergeant (PSG), Sergeant First Class (SFC) Frank G... and I was scared shitless!

In the B Company area, all of the platoon offices — if you could call it that — were housed in one room, a room not much bigger than a good-sized bedroom of a single family home. There were three desks in the room — one at each corner — and in that room when I walked in with the Company Commander were three Non-Commissioned Officers (NCOs) — Staff Sergeant (SSG) Joe E. who was the Platoon Leader for one of the platoons, more on him later; SFC B., the Platoon Sergeant for my good friend, Ken Golla (d. March 1994), who was tragically killed at Green Ramp at Ft. Bragg a few years later; and SFC Frank G., my Platoon Sergeant. Ken was the only officer platoon leader in the company at the time. I would be the second.

I don't remember much of that initial interaction with Frank, but what I do recall is that he looked about 20 feet tall and intimidating as hell. In retrospect, he was very welcoming, but he made it clear that my job was not to be hanging out in the office doing paperwork, but to be out with the platoon.

The first couple of days, I followed him around like a puppy-dog and hung on every word and piece of advice he shared. Finally, after about a week, he asked me if I was going to counsel him. I literally laughed. I said something to the effect of, "I thought you were going to counsel me." You see, at West Point and IOBC, they made it sound so easy — sit down with your PSG for his initial counseling, lay out your expectations, blah, blah, blah. What they fail to make you realize or consider is that you're an officer with less than two years in the Army and you're about to tell this 8-10 year Army senior NCO how it's done and what you expect from him. In hindsight, maybe they did tell me that, but I sure as hell didn't remember!

So what happened? Frank and I went to the Burger Bar (real name) on Panzer Kaserne for lunch and we had a professional discussion. He told me what he expected of me as his Platoon Leader (PL); I told him what I expected of him as my PSG; we talked about how we would lead the platoon as a command team; and that as an officer, my decision was final, even if he disagreed with me, and that he would support the decision.

This had a huge impact on me and shaped how I would lead units throughout the rest of my career. My senior NCO and I were a command team, and we worked hard to make each other better, because it was the right thing to do for the Soldiers, and the unit. I struck gold as a young officer with exceptional NCOs right off the bat. If it went the other way, and I didn't have the gift of excellent counsel and advice from the NCOs in B Company, there's no telling how it would have worked out.

When I joined the Army and reported to West Point for Beast Barracks (aka Cadet Basic Training) in the summer of 1985, Soldiers who fought in Vietnam were still serving in the operational Army. One such Soldier was SSG Joe E. As mentioned, we met in 1990 when I reported to 1-16 IN. Joe was a tunnel rat in Vietnam. As he explained to me, his size and stature made him the perfect Soldier to enter and clear tunnels in Vietnam.

I was blown away by some of his stories and lessons. Joe had a break in service after the war, and then he later rejoined the Army. He shared sage advice with me one day via his three principles that in his opinion defined a good lieutenant (LT) — lessons that I extrapolated out for the remainder of my career:

1. A LT that will listen to his NCOs when they are right.
2. A LT that will allow his PSG to accomplish his missions when in garrison.
3. A LT that knows when to step in and drop the fucking hammer.

One other thing that Joe shared with me that I have never forgotten and that I share with leaders all the time: Your troops are most always an accurate reflection of yourself. If you lead with standards, Soldiers will follow with standards. It's funny — my Battalion Command Sergeant Major (CSM) would say, and live, the same thing many years later.

GETTING LOST

In the days before Global Positioning Systems, Blue Force Tracker, FBCB2, or any other high-speed navigational device the Army threw at us, you needed a "100 mph finger and your head on a swivel" to navigate from a moving Bradley Fighting Vehicle. In the summer of 1990, I found myself at the Combat Maneuver Training Center (CMTC) in Hohenfels, Germany. My platoon was one small part of our brigade's annual training rotation — about 10 days in the "maneuver box" conducting multiple combat operations against a living, thinking opposing force (OPFOR).

There was no train-up. This was literally the first time I would maneuver my platoon in a training exercise, and at the time, CMTC was "the show". Urban legend had it that you could get fired if you sucked... or maybe it was real. Either way, I didn't want to screw this up.

Two significant events occurred for me on this rotation: first, I took a wrong turn on the battalion attack and got lost; and second, I learned how to play spades. The latter would serve me well in talking shit and holding my own with my platoon on the internal battlefield of unit card games. The former was no doubt the more professionally developing experience.

It was the Brigade attack... the super bowl event of the rotation. The Iron Rangers were the lead battalion. B Company would be the lead company. For whatever reason, my company commander selected my platoon to be the lead platoon... lead platoon of the lead company of the lead battalion for the brigade attack. What could go wrong? I can think of no reasonable explanation why CPT Doug B. picked my platoon other than he thought that there was no way I would be the consummate dumb ass new platoon leader and actually lead the formation myself. Of course, that's exactly what I did.

I do remember that after receiving the company operations order (OPORD) and returning to my platoon, I pulled my PSG aside, SFC Frank G., and told him that we were leading this bitch! I was stoked! I'm not sure exactly how it went down 30+ years ago, but I feel like Frank put his hand on my shoulder and said something like, "calm down, little fella."

Regardless of how it went down, when I issued my platoon OPORD, I had my Bradley as first in the order of march. My squad leaders and PSG disagreed with this and recommended that I place my Bradley as second in the order of march.

There were several reasons for their recommendation. First, and most obvious, I was a rookie and had never navigated my Bradley on the move (outside of the Bradley Leader Course... which was really — never!), much less as part of a Brigade exercise and wearing night vision goggles (NVGs), first generation NVGs at that.

It would be an early morning attack — i.e., it was very, very dark. Second, and more tactically sound, was that the best place to command and control (C2) my platoon was as second in the order of march. With my best squad leader in the lead, he would take care of the navigation, allowing me to C2 my platoon, call for fire as needed, make decisions, keep my boss informed, etc. It all made total sense. Of course I listened to my NCOs... not! I took the lead. Don't worry... I got this!

The rest, as they say, is history. In the simplest of terms, we crossed the line of departure for the attack with me squarely in the lead. Overwhelmed, confused, and generally a mess, I managed to keep us on track for a while until I came to the literal crossroads, and I went right when I should have gone left. With nothing but my map board, NVGs, and terrain association, I was lost in the sauce. My entire platoon followed me. Fortunately, the follow-on platoon didn't follow us and ended up keeping the company, battalion, and by extension, the brigade on track and got them to the fight.

Eventually, I got my platoon turned around and we got into the fight. Lesson learned — listen to your NCOs! SSG Joe E.'s words were ringing in my ears. And also... be humble enough to admit you made a mistake. I took some shit that day, and for good reason, but I learned some valuable lessons. Development never ends.

EIB — FAILING MY ETHICAL CHALLENGE

The Expert Infantryman Badge (EIB) is THE mark of an Infantryman. In 1991 when I tested for it, the EIB was a huge deal. It still is today, but I suppose some of this excellence has been watered down by the multiple, different badges that the Army awards for "excellence." In the early 1990s, the only two badges that I recall which were awarded for excellence in one's Military Occupational Specialty, or MOS, were the EIB and the Expert Field Medical Badge (EFMB). So, yeah...this was a big deal and for a brand-new Infantry Lieutenant, I really needed to earn my EIB the first time around. Spoiler alert. I earned my EIB. In fact, I didn't just earn it, I was "true blue," which means I did not receive a single no-go. But I did fail... big time.

At the time, the EIB test consisted of roughly 30-some individual Infantry tasks like call for fire; assemble and disassemble a .50 caliber machine gun; install and employ a claymore mine; administer first aid; complete a 12-mile foot march in three hours; etc. There was also a Physical Fitness test, and you obviously had to fire as an expert with your assigned weapon, which at the time was an M16. Over the years the tasks have changed slightly, but the test has generally remained the same. One constant has been the land navigation requirement — both day and night land navigation.

For day land navigation, you had to find a certain number of points in an allotted time, using a map, compass, pace count, and terrain association. You could not use roads, and you could not talk with anyone on the course. For night land navigation, it was a similar requirement — a certain number of points in a designated time frame. The main difference was that you only had a compass — no map. It was very much a test of your ability to use a compass and pace count. We called it "dead reckoning."

There were no other navigational aids, no night vision devices, and it was an individual event; again, meaning that you couldn't use roads and you could not talk to anyone on the course... and it was dark... really dark.

For the most part, I never really struggled with land navigation (with the obvious exception of the previously told

CMTC fiasco!). I had no trouble during the EIB either — I smoked both day and night land navigation... but again, I failed.

I graduated from the United States Military Academy in 1989. For four years I lived under the honor code that a "cadet will not lie, cheat, or steal, or tolerate those who do." It was part of the fabric of West Point, who we were, and immortalized in our motto, "Duty, Honor, Country." When I graduated and was commissioned as a Second Lieutenant (2LT) of Infantry, one would think that I'd be more than prepared to make the right decision and do the right thing when faced with an ethical dilemma. I was not.

The EIB test is intentionally difficult and at the time, night land navigation was one of the last events. By the time we got to this event, there were only about 5-10 of us left from my battalion (it was a division-run EIB). A few hours before we were to begin night land navigation, I was in a General Purpose (GP) Medium tent with the remainder of the guys from my battalion. In this tent designed to hold about 30 Soldiers, at most there were 10 of us. Most of them were gathered in one corner of the tent. I was down at the opposite end, readying my gear and basically hanging out.

There were whispering and hushed voices coming from the area where the small group assembled, as well as the sound of the rustling and cutting of paper maps. It didn't take a rocket scientist to figure out what was going on — they had a cheat sheet. More specifically, someone had gotten ahold of a map with all the points and these guys were transcribing those points onto a pocket-sized map that they would carry with them on the course that night.

Again, it's dead reckoning — just a compass and your pace count — no maps. I recall vividly what they were doing because I watched this go on from my position at the other end of the tent. I may have even wandered over there to get a better look. I also recall that it was Soldiers of all ranks, to include NCOs and officers. None of them were very concerned about me doing anything about it, and I'm sure they extended the invite to me to get a look at the points, but I can't remember for sure.

So, what did I do? I did nothing. I went out that night and took the test as prescribed and I passed. But I failed. For many, many years I never shared this story with anyone because I was so incredibly disappointed in myself. Here I was, a West Point graduate and an Officer in the United States Army, and even though I passed the event, I failed myself and my teammates. What should I have done? Maybe I should have turned them in; maybe I should have confronted them and told them to get rid of the cheat sheet/map; maybe I should have lectured them about what they were doing was wrong and let them make their own decision. But when push came to shove, I did nothing.

I started telling this story more often toward the end of my career. When I was stationed at West Point, I used it as a lesson for the cadets about honor and integrity, and to highlight that what we think may be black and white in the classroom is more often gray in the real world. And that integrity comes with intestinal fortitude, choosing the harder right over the easier wrong. I would like to think that this experience made me a better leader and officer over the course of my career.

SUPPORT PLATOON — I WANTED THE SCOUTS

There is no doubt that the coolest and sexiest job for any lieutenant (LT) in an Infantry battalion is the Scout Platoon Leader (PL) gig. To be the Scout PL is hands down the position all of us self-anointed high speed LTs wanted to be. About once every year to 18 months, the battalion commander looks at moving LTs to other jobs for professional development or to fill critical positions or shortages. The normal turnover is accompanied by the arrival and departure of new officers or more senior officers, respectively.

Later in my life as a Battalion Commander myself, and now the person responsible for managing the lieutenant slate, I tried to keep my LTs in platoon leader positions for as long as possible. Short of incompetence or the inability to lead, the more time a young officer could spend as a rifle platoon leader, the more developmental it was for him. Leading a rifle platoon is where the rubber meets the road and I believe in the long term there was more growth and development leading a platoon of 30+ Infantrymen than making slides in the S3 shop. I'd rather go without that Assistant S3 than pulling an officer pre-maturely from a platoon.

Likewise, when I was moving LTs to a specialty platoon — Battalion Scouts, Battalion Mortars, Battalion Support Platoon — I considered expertise, character, maturity, and other intangible qualities like communication, problem solving, and decision making. It wasn't always about the best lieutenant — it was about the right lieutenant. And in extremely rare cases — like once — would I put a brand new 2LT into one of these specialty platoons (and it was the mortar platoon which requires some specialty training via the Infantry Mortar Leader's Course at Ft. Benning.)

Which brings us back to 1LT Reed, a fairly senior lieutenant in the battalion with about 1.5 years as a rifle platoon leader. I was a pretty good LT and felt that I would have an opportunity for a specialty platoon, and because I was a legend in my own mind, I wanted the Scouts. Quick sidebar — you also need senior LTs to serve as company executive officers. Again, in my later life as a battalion commander, I wanted to put the right officer into these positions, but most importantly, they needed to be able to command the company in the absence of the company commander.

I was a Battalion Commander (BC) from 2006-2009, at the height of the Global War on Terror, so this was a very real concern. You were literally one bullet away from commanding that company of approximately 130 Infantry Soldiers. Officer assignments took time and patience and needed to be informed by what was first... best for the unit, and then the officer, second.

OK... back to 1LT Reed. As mentioned, I wanted the Scouts... instead, I became the Support Platoon Leader. The greatness of my battalion commander at the time, Lieutenant Colonel (LTC) Ric Crosby (d. 2008), was that he cared about us, so when it came time for the LT moves, he didn't just push out some memorandum with his signature on it and we all moved out to our new jobs. No, instead he talked to each one of us. He did this frequently as part of his junior officer mentorship and counseling program, something I adopted in my career.

When LTC Crosby called me into his office and told me I was going to take over the support platoon, my face said it all, to which he commented, "So you wanted the scouts." At the time, I wasn't smart enough to say something like, "Sir, I'll do whatever is best for the battalion."

Instead, I said, "Yes, sir, I did." And instead of telling me to shut up and move out, LTC Crosby explained to me that the Support Platoon is the most important platoon in a mechanized Infantry battalion — responsible for all classes of supply — chow, fuel, barrier material, ammunition — and that the support platoon leader is also the Mess Hall officer, responsible for the daily operations of the installation dining facility (little did I know at the time that the DFAC would be the source of the best ass chewing I ever received — from the Brigade Commander). In short, the Support Platoon touches everything that the Battalion does.

He also went on to explain that he needed strong leadership in the platoon given the complexity of the multiple branch specialties — fuelers, cooks, Infantrymen (who didn't necessarily want to be there), truck drivers, ammunition handlers, HAZMAT specialists, etc. And, by the way, there was an ongoing drug trafficking investigation involving multiple Soldiers in the platoon. In fact, the introduction to my platoon at my first formation involved the arrest of two of my Soldiers!

The good news was that he was going to give me the best platoon sergeant in the battalion. I wasn't sure if I was on the receiving end of the best information operations campaign ever, or if he was being serious. He was serious. I was humbled and soon realized that the Battalion Commander handed me an incredible leadership opportunity.

Of course, LTC Crosby didn't owe me any kind of explanation. But in that 10-minute conversation, he taught me more about leadership than anything I ever learned in the classroom. And so, SFC M. and I led the support platoon and navigated multiple training events and leadership "opportunities." In short, I became a better officer and leader because of this experience.

My next, and last lieutenant gig, was as a Company Executive Officer (XO). I felt very prepared for that job, given the professional experiences I had to that point. I went to Alpha Company, and I was the XO for A/2-6 IN for about eight months. It was truly a great job, with a great boss, and it prepared me for the next step, which would be Infantry company command. CPT Eric N. was a fantastic mentor. But I'm not going to lie. The job was a relative cakewalk after being the Support Platoon Leader for 2nd Battalion, 6th Infantry Regiment at Ferris Barracks, Germany. LTC Crosby knew what he was doing! Thank you, Sir, for the opportunity. I am better because of it. Regulars on Guard!

3. LEARNING TO BE AN INFANTRY OFFICER

THE RSOP FAIL, AIR ASSAULT, AND THE DIVISION EDRE

The Readiness Standard Operating Procedure, or RSOP, is the go-to-manual for how units do business upon notification and subsequent deployment. Processes, systems, and how we deploy formations have changed over the years, but back in 1993 when I signed-in to the Bronco Brigade — 3rd Brigade, 25th Infantry Division at Schofield Barracks, Hawaii — the Brigade RSOP was the gospel.

As a brand-new Captain coming off one overseas assignment (Germany) and heading to another (Hawaii) — by way of six months at Ft. Benning, GA for the Infantry Officer Advanced Course — I was pumped to have the opportunity to serve in the 25th Infantry Division. I very much wanted to command a light Infantry company and my wife and I (and our growing family) were excited to experience Hawaii. And the 25th ID was light... very light. Our primary means of movement was on foot via our LPCs — Leather Personnel Carriers!

Before taking command, however, my first job was to serve as an Assistant Brigade Operations Officer, or Brigade S3 Air, responsible for all air and deployment operations in the Bronco Brigade. Coming from a pure heavy/mechanized Infantry assignment as a Lieutenant, I was probably the least prepared person for this job. But there was no Plan B... I needed to figure it out.

My immediate task was the rewrite of the Brigade RSOP. I inherited a decent product from my predecessor, but it needed some work. Leveraging the expertise of my Air NCO and talking to others throughout the Division and Brigade, I put together what I thought was a pretty good product. The Brigade just went through a Change of Command, and COL Gary S. was the new Brigade Commander, Bronco 6. I needed to get the revised RSOP to him for signature. And almost back-to-back, within a couple of months after the change of command, two Division-level exercises were scheduled which would teach me more than any classroom experience I ever had.

First, the RSOP. I printed out the cover page of the RSOP which was a document outlining the purpose and content of the RSOP, establishing it as Brigade policy, and it had Bronco 6's signature block. I put it in a folder with a routing note and gave it to the Brigade Adjutant to get it to the boss. The plan — my plan — was to have the Brigade Commander sign the cover memorandum, slap it on top of the 100+ page RSOP, and then take it to the print plant for distribution. Remember, this is 1993. Word processing and print production were an ordeal inside Infantry brigades at the time.

Instead, I got a call from the adjutant about a day later summoning me to Bronco 6's office. I went in, reported, and stood at attention. The boss asked me where the hard copy of the RSOP was located. I said in my office — the one and only copy. He asked me if I expected him to sign the cover memorandum approving the RSOP, thereby making this Brigade policy and standard operating procedure (SOP), without reading the actual RSOP itself.

Of course, I did... but there was no way I was saying that. I immediately recognized what a dumb ass I was. COL S. told me to go get the printed RSOP and bring it to his office. After he read it, he'd decide on whether to sign or not.

Several days later, I got back the RSOP with Bronco 6's edits. I feel like it was covered in red ink... but it may have been green. Either way, there were a lot of comments, edits, and follow-ups, to include that the document was not in compliance with the Army's manual for preparing and managing correspondence, Army Regulation (AR) 25-50.

And again, this was 1993, so it wasn't quite as easy as going back to my saved document on my Dell computer and knocking out these changes, especially since I didn't have one! I look back on this experience and think about what a jackass I was... of course the Brigade Commander would approve a document without even laying eyes on it... how could I be such an idiot.

I never forgot this experience and later in my career when I was a Battalion and then Brigade Commander, I read everything my staff put in front of me. And, the RSOP was eventually signed, printed, and distributed to the Brigade.

Within those first three months in Hawaii, I found myself not only trying to make sense of the RSOP and learn my job, but now planning a Brigade Air Assault and Ground Movement to the Kahukus and Makua Valley, two of our main training areas on Oahu.

Looking back on this, what an experience. We did battalion air assaults, live fires, truck movements, foot marches from one end of the island to the other. Incredible. At the time, though, I seriously doubted my abilities, but like I said before, there was no Plan B... so I figured it out. The biggest lesson I learned during that experience had nothing to do with the planning and execution of this incredible training event. Rather, it was that as a light fighter I needed to "travel light, freeze at night."

Remember, I was a mechanized Infantryman before arriving in Hawaii. In the mechanized world, there was no limit on what you could take to the field... as long as it fit in your Bradley. Coolers of soda, pogey-bait galore, cold weather gear, creature comforts, the kitchen sink — there was no need to be miserable.

So, when it came time to pack my gear for this huge exercise in Hawaii, I was in a bit of a conundrum. I didn't want to look like a complete fool, so I asked around about how to pack my rucksack, what to bring, etc. — kind of like I was asking for a friend.

I also knew that I would have a High Mobility Multipurpose Wheeled Vehicle (HMMWV — or, for the uninitiated, kind of like a Jeep) which was a rarity in light formations.

I showed up at the office that morning ready to go to the field. I drug my rucksack up to the 3rd floor... completely overloaded with clothes, food, and other things I might need, complete with a brand-new Army-issue puss pad strapped to the top... and then I went back to my car and brought up my sleeping bag, tightly rolled and in a wet weather bag.

My Air NCO looked at my mess and asked me where I was going. I said, "To the field." Needless to say, I got a lot of help from the other NCOs and officers in the S3 shop — and by help I mean they totally busted my balls! You need a thick skin in the Army. If you can't take a joke or handle the banter and sarcasm, then it is going to be a short career!

Fortunately, there was no social media at the time to show off my stupidity. The sleeping bag and puss pad got tossed; the extra pair of boots... tossed; the extra uniform... tossed (we used 100mph tape to repair torn crotches and knees in the field).

In went extra water and MREs... out went the twinkies and pringles. Out went the shelter half and poles; in went the poncho and bungees for my poncho hooch; an extra poncho for a ground mat; and one poncho liner... not two! And in went extra socks. "Travel light, freeze at night" — take mission essential equipment; what you need to fight and win and sustain yourself — Soldier's load matters.

Shortly after this, I took part in another professional development exercise. An Emergency Deployment Readiness Exercise, or EDRE, is designed to be a no-notice alert and test of a unit's deployment readiness. In the summer of 1993, depending on one's perspective, I had the fortunate — or unfortunate — opportunity to be at the center of a Brigade EDRE. This, of course, was all on the heels of my RSOP fiasco which resulted in some great lessons learned, and a pretty good document, and my first major exercise as a brand new staff officer, and S3 Air, in the 25th Infantry Division.

I was definitely feeling more confident in my abilities, but in no way did I feel confident enough to manage a Brigade-level deployment. In short, this is like Game 7 of the World Series as the Brigade S3 Air... and I did not want to be on the losing side.

In reality, a no-notice exercise should not be a big deal. If a unit has sound SOPs and good processes in place, it's just a matter of execution. I worked hard to make sure we had these systems, both at the Brigade-level and down to the Battalions and companies. This required coordination and iterative communication with my counterparts in the six battalions across the Brigade Combat Team (BCT), as well as constant dialogue and relationship building with my counterparts at Division.

And, of course, because any deployment would be completed on the back of our Air Force brothers and sisters, I needed to maintain good relationships with the air operations folks, loadmasters, etc. And finally, internal drills and rehearsals were critical to maintain proficiency and competency.

There are a few things I remember when we executed the EDRE. Although I had good intelligence, or early warning, from my contacts on when notification would occur and when we would begin the 18-hour deployment sequence, I was still caught off guard. We ended up moving our vehicles and personnel from Schofield Barracks to Hickam Air Force Base where C-141s, C-5s, and C-130s transported the BCT to the Big Island and Pohakuloa Training Area (PTA) for a Brigade-level exercise. Our processes worked and this experience later served me well as a Battalion Executive Officer when I deployed 1-22 IN to Iraq in 2003. But that's for later.

Perhaps, though, the most dramatic thing I remember about that experience was sitting in the jump seat in the cockpit of a C-130 on the last bird out of Hickam. As we were ready to land, I remember thinking how cool this was... there I was in the front row, practically flying the plane. This feeling of euphoria lasted right up until we touched down, blew out a tire, and skidded to a stop. It was at that point I realized I chose the right service — the Army! I was glad that I was a light Infantryman and from now on I would always be seated in the back of any Air Force aircraft. The cockpit was not the place for me!

COMPANY COMMAND(S) AND MY FIRST DEPLOYMENT

After cutting my teeth as a Brigade staff officer for about a year, I was finally slated to take command of C Company, 4th Battalion, 87th Infantry. I don't think it is an overgeneralization to say that Infantry company command is what we all aspire to when we are commissioned as 2LTs of Infantry coming out of USMA, ROTC, or OCS. The moment was finally here... and then it wasn't.

A few weeks prior to beginning my change of command inventory, the Brigade was notified of a pending deployment in support of Operation Uphold Democracy in Haiti. I won't recount the history of the operation here, but this was the planned invasion of Haiti to restore President Aristide to power after a military coup unseated him. At the final hour, diplomatic negotiations resulted in Aristide's restoration to the presidency and the invasion became a peacekeeping and stability operation. So, in early January 1995, 3rd Brigade prepared to deploy to northern Haiti, in the vicinity of Cap Haitien.

Since I was the Brigade S3 Air and my core mission was to deploy the Brigade, my change of command was delayed until we successfully had the entire formation deployed and had completed the transfer of authority and relief in place with the Marines. While no doubt disappointing, this made complete sense, and it was cool to take command of a rifle company that was operationally deployed. Within hours of the change of command, we almost immediately moved out on a several day out of sector mission to actively patrol a series of towns and villages outside the Brigade's normal area of operations, and provide security for the local population.

A couple of things stand out about that deployment. First, I was blessed with an exceptional 1SG in the person of Art L. An incredible Infantryman, leader, and friend, he was the type of NCO who modeled the first sentence of the NCO Creed, "No one is more professional than I."

Art taught me how to be a company commander, how to lead in an operational environment, how to trust my platoon leaders and platoon sergeants to accomplish my intent, and how to be a good team player with my fellow company commanders. The other thing about Art was that we laughed... a lot. There is a lot of stress associated with a deployment, but there also is a lot of funny shit that happens. It's good to laugh! He taught me that it's OK to have fun and laugh, and to especially laugh at myself. While never a problem for me, Art made it normal to have fun!

Art also taught me how to manage the intersection between the battalion commander and the battalion staff. This is a whole new experience for a company commander. You work for the battalion commander... period. That said, some Battalion Commanders lead through their staff, and others lead through their company commanders.

I commanded two rifle companies in two different battalions, with two totally different battalion commanders. Each had their own command philosophy and leadership approach based on their personalities, style, experience, etc. Art taught me how and when to keep my boss informed, what I could blow off from the staff (sorry, it's true), what I couldn't, and what ass-chewings were worth taking, and which ones weren't. There is no book or doctrinal manual that will teach you this. I was lucky to have Art L. as my 1SG.

Second, I worked for a Brigade Commander who was exacting, precise, standards based, and the consummate professional. I loved Bronco 6, even though I'm sure I drove him nuts at times — starting with the RSOP fiasco — and there were a handful of times I really questioned some of what he was telling me. Good thing I listened! It was from him that I learned how to be an Infantry Officer and Leader.

During Operation Uphold Democracy, we worked closely with Special Forces Operational Detachments — Alpha (ODAs). This would often mean that you had two units — my rifle company and an ODA — conducting tactical operations in the same operational space. Both organizations were commanded by officers of the same rank — Captains. Both units had similar missions — providing security and stability in a specified area.

As the overall "battlespace owner," Bronco 6 would require a joint operations brief from the two commanders prior to any mission that involved one of his rifle companies and an ODA. While these types of operations were not everyday events, they happened a lot. My little pea-brained captain mind thought that this was sort of micromanaging and "didn't Bronco 6 have better things to do." Needless to say, I was still learning!

And what I did learn was how to coordinate operations with another unit; how to work together to build an intelligence picture; how to communicate across radio platforms that were not always compatible; and how to organize logistics and medical support. And I also learned that there can only be one commander, and as the battlespace owner it was Bronco 6, and by extension, me, the company commander.

In short, it was Ph.D. level stuff. The briefs we put together were the forefathers of what we would later do regularly during the operations in the Global War on Terror (GWOT). All of what I learned from Bronco 6 would serve me well as I progressed throughout my career.

Shortly after our re-deployment from Haiti, the Army went through a re-organization. While these decisions were way above my pay grade, I was on the receiving end. My Battalion, 4-87 IN, was going away. I had only been in command for about four to five months. I needed another command to be what the Army termed "branch qualified" — in short, I needed at least 12 total months of company command.

Fortunately for me, Bronco 6 gave me another opportunity and I soon took command of Alpha Company, 2nd Battalion, 27th Infantry — the Wolfhounds!

Having the opportunity to command two rifle companies is not very common, so this was truly a blessing. Equally awesome was that I was joining a great team with some great peers, a great command climate, and a unit with an incredible history.

It was with A Company that I first recognized the power of being part of an organization with a long history and lineage. It was the first time I thought about being part of something much larger than myself — we were not just representing A/2-27 IN... we were representing all who went before us.

Desiderio, Millet, Foley — all were former company commanders of A/2-27 (in the case of Desiderio and Millet the unit was designated as E/27th IN, or Easy Company, 27th Infantry Regiment) — and all were awarded the Congressional Medal of Honor, the military's highest decoration for valor in combat.

And then there is Reed... the current A Company Commander... stepping into this legacy of greatness. There was no pressure at all — ha!

Shortly after taking the guidon and command of the Company, the 25th Infantry Division was preparing to host a celebration to honor the 50th Anniversary of the End of the War in the Pacific. The events welcomed back and celebrated veterans who served in World War II. One of those members was COL (R) Lew Millet, who was also the Honorary Colonel of the 27th Infantry Regiment. Lew was a three-war veteran. He fought in World War II, Korea, and Vietnam. It was in Korea, as the commander of Easy Company, 27th Infantry (present day A/2-27), where he earned the Medal of Honor.

He took command of the company after the previous commander, Captain (CPT) Reginald B. Desiderio was killed in action — an action for which he would posthumously be awarded the Medal of Honor. As Lew said to me, "Try taking over command for a guy who earned the Medal of Honor."

It was a good perspective. But think about this for a minute. Two company commanders of Easy Company earned the Medal of Honor back-to-back during the Korean War. Talk about leading from the front and setting an example for your Soldiers. Incredible. This is inspired leadership in action.

Lew spent about a week with us in Hawaii. Even though he had obligations at the state, city, division, and regimental level, he spent his most time with us, the Soldiers of Alpha Company, 2nd Battalion 27th Infantry. It was his company. The company he led in combat; the company that in February 1951 followed him up Hill 180 with fixed bayonets; the company that rallied around him until the day he died in 2009.

There was no way the men of present day A Company were going to let him down. COL (R) Millet's visit and time with the company was a transformative experience for me and for the rest of the Soldiers. Being part of A Company was something bigger than ourselves. We represented — we were a part of — a legacy of toughness, excellence, and determination. We were the warrior class.

We called Hawaii the "land of the lost" because during training exercises and foot movements you could disappear into some of the gulches — basically enormous valleys — and enter a whole new world. Navigation was done the old-fashioned way — with a map and compass. Each company had one or two first generation GPS systems (called a "plugger") but they weren't conducive to navigating on the move, and they were heavy... and required batteries.

As a light Infantry formation, a Soldier's load mattered, so we navigated the way Infantrymen have done for years — a map and compass. What you carried and how you moved and used the terrain was an art that was forged through numerous training exercises throughout the Hawaiian Islands.

My time as a light Infantry company commander refined

the basics for me — the essence of leading Soldiers, maneuvering a light Infantry company, employing fire support, marksmanship, physical fitness, resupply. I was very fortunate to serve as a company commander before the technology took off.

Don't get me wrong. I'm a huge proponent of leveraging technology and all the systems at your disposal. But it still comes down to blocking and tackling. The basics matter and people matter, and it was in the 25th Infantry Division where this was really ingrained into me, as well as how to train, how to live fire, how to maintain accountability of my Soldiers and equipment... how to be a warfighter.

4. BEING A FIELD GRADE

HEADING TO 4TH INFANTRY DIVISION...AND A LESSON ON BEING VALUE-ADDED

After my time in Hawaii, the Army afforded me an awesome opportunity to earn my Master of Arts in Sociology from the University of Maryland; followed by an assignment on the faculty at the United States Military Academy (USMA). This took me out of the operational force for four years (two years of grad school and two years at USMA). I was ready to get back into the mix. After a year at the Command and General Staff College (CGSC) at Ft. Leavenworth, KS., I reported to the 4th Infantry Division at Ft. Hood, Texas in June 2001. It's funny, though... because I didn't want to go to Hood!

Coming out of Leavenworth and CGSC, I wanted to go to the 10th Mountain Division, and specifically to the Polar Bears of 4th Battalion, 31st Infantry Regiment. I very much wanted to go back to a Light Infantry Division and the 10th Mountain Division and Ft. Drum is where I wanted to be.

I interviewed with the Battalion Commander who made me his first choice. And at that time, not many people were clamoring to get to Watertown, NY where Ft. Drum was located. I thought I was good to go. Years later, as a Brigade

Commander who made these hiring decisions, I laugh about how naive MAJ Reed was at that time!

When the assignments came out, I was going to the 1st Cavalry Division — First Team — and Ft. Hood, and instead of going right down to an Infantry battalion, I was going to be on the Division staff. Disappointed would be an understatement. And for the first time in my career (and definitely not the last), I leveraged my professional network and my mentor, then-Colonel (COL) Bob Cone. [GEN Cone would later become the TRADOC Commanding General. He passed away in September 2016 from cancer.]

When he was a Captain, COL Cone taught me *PL300: Military Leadership* as a junior (cow) at USMA. That was the beginning of a professional and personal relationship that would last until he died.

COL Cone was the Commander of 2nd Brigade, 4th Infantry Division at Hood. I called him up and explained my situation. I was coming to Hood; could he get me to the 4th Infantry Division; could he get me to his Brigade; could I get to a battalion as an operations officer (S3) or executive officer (XO)? As I write this, I feel like I made a lot of demands to him! True to form, he went to the mat for me. However, as the musical philosopher, Meatloaf, so eloquently puts it: "Two out of three ain't bad."

My assignment was changed to 4th Infantry Division and I was going to be the S3 for 1st Battalion, 22nd Infantry Regiment (Regulars) in 1st Brigade, not 2nd Brigade — not COL Cone's Brigade. To this day, I'm not really sure why, but the trajectory of my career would be forever changed because of this. Little did I know at the time what this would mean for me as a Professional Soldier.

And so off I went, and I reported to 1-22 IN (Deeds not Words) in June 2001. From 2001 to 2004 I would have the trifecta of Field Grade/Major jobs: Battalion S3, 1-22 IN; Battalion XO, 1-22 IN; Brigade S3, 1st Brigade. I couldn't have asked for a better opportunity, but it about crushed me! Majors run battalions and brigades. The term "Iron Major" is a real thing!

One quick sidebar about then COL Cone. During Physical Training (PT) hours, if you ran past his Brigade Headquarters building on Battalion Avenue at Ft. Hood, you would hear Johnny Cash's classic song, *Ring of Fire*, blaring from the loudspeakers atop the building. Even though my Brigade's designated PT space was further down the road, I would frequently detour past 2nd Brigade to hear the Country Music legend belt out his classic song on repeat!

Years later, when I became a Brigade Commander, in honor of General (GEN) Cone, I would have my team play *Ring of Fire* at the beginning and the end of every quarterly Brigade Run. It was my way of honoring the legend who is General Bob Cone. Rest in Peace, Sir!

In August of 2001, after about two months in the seat as the battalion operations officer, I found myself in the motor pool, which was an unusual place for the S3 to be. This is usually the domain of the Battalion Executive Officer (XO). I'm not sure why I was there, but there I was. During the course of my motor pool walkabout, I came up on a group of young LTs talking and laughing. I startled them at first. As I got closer to the group, they immediately stopped talking and looked awkwardly at the ground, in the air, off in the distance... anywhere but looking at me. Eventually, one of them locked eyes with me, called the group to attention, and saluted me with the greeting of the day.

I returned the salute and then asked them what they were talking about. This was met with silence, and everyone was uncomfortable. There was no way I was going to let this go... they were clearly talking about me! So, I repeated the question and after no response and even more uncomfortableness, I said something to the effect: "OK, I get it, you're talking about me. I hope it's good."

One brave LT finally spoke up and said, "Sir, we weren't talking about you. We were talking about CPT X. He's a seagull." Of course, I had to follow-up, so I asked, "What's a seagull?" The LT looked around and gathered his thoughts. "You know, sir, a seagull... he flies in, shits all over the place, and then flies out."

After stifling a laugh, I said something profound like, "Carry on," and I walked away, not sure what to say. I probably should have tightened them up for being disrespectful of a ranking officer. But you know what, they were right. The lesson here is to be value-added. As I became a more senior leader, this was a great lesson to remember as I would go and visit training, ranges, or attend briefings given by more junior leaders.

Catch people doing the right thing. Let the visit be about making people better. Leave the right impression; leave a positive impression. Make those around you better. Be value-added. Don't be a seagull!

9/11 - GTMO - BUILD UP TO INVASION

When I arrived at 4th ID, the Division was just coming out of a several year phase where they were the Army's Force XXI "experimental" division. As the Army's main effort for modernization, all sorts of new systems, technology, technical advances, and weapon systems were fielded and tested with the unit.

Then Major General (MG) Ray Odierno took command of the division shortly after I arrived, removed all signage that referred to us as "experimental" and pronounced the division ready to deploy, fight, and win. The timing of this was oddly fortuitous in the sense that you now have a division coming out of this period of testing and experimentation and ready to deploy... and then 9/11 happens.

Like many others, I remember exactly where I was that morning. We had just finished physical training, and I had showered and changed at the post gym and returned to my office. The Battalion Intelligence Officer (S2) came into my office and told me that I should come into our operations center to take a look at what was going on in New York. Our Brigade (the Raider Brigade — 1st Brigade) was preparing to assume the Division Ready Brigade 1 (DRB1) at the end of the week; and our Battalion (1-22 IN) was going to be the Division Ready Force 1 (DRF1).

For the uninitiated, this essentially means that we were going to be on an 18-hour recall to deploy anywhere in the world on short notice. At the time, there were different types of DRB/DRF packages in the Army — light, airborne, and heavy. We were the designated heavy, or mechanized, formation. As part of our preparation, we converted our battalion conference room into an operations center, complete with radios, digital communications, maps, etc... and televisions so we could maintain a 24 hour news cycle.

So, it was in this conference room/operations center that I found myself at approximately 0800 Central Time, looking at one of these TVs, as the second plane crashed into the World Trade Center. As I write this, I can feel the same goosebumps and anxiety I felt that morning; the same "WTF" moment. What followed that day at Ft. Hood is similar to what happened at almost every other Army installation — we locked down the post and prepared for war. Nothing was ever the same again.

After getting through the first several weeks of the new normal, we began to settle back into a routine, or at least some semblance of a routine. At the time, 4th ID ran a "red, amber, green" cycle, commonly known as the RAG. As I recall, each cycle lasted roughly six-weeks. Red was the tasking cycle. A brigade and/or a battalion in the red cycle performed all assigned tasks and duties — like guard — that the Division designated. The saying went: "red 'till you're dead." Green was the prime-time training cycle. If a unit was green, they had priority of ranges, land, and resources. When 9/11 happened, we were in the amber, or mission, cycle and we immediately expected to deploy. We did not and instead remained postured and ready to go for the remainder of the amber cycle.

As one might imagine, the rumors were out of control... frankly, they became a bigger problem than any Soldier shenanigans we might have been dealing with at that time. Of course, everyone in the battalion was amped up and ready to get in the fight. At the time, Afghanistan — or Operation Enduring Freedom — was the only show in town and it began to look more and more like 4ID wasn't going to be in that fight. Until one day we were... sort of.

I don't recall the exact timing or really how all of this materialized, but in early 2002 a mission seemingly appeared out of nowhere to support Joint Task Force (JTF) 160 in Guantanamo Bay (GTMO), Cuba.

How it ended up with 1-22 IN is beyond me, but it obviously came through the Division, to the Brigade, and then to the Battalion. As the Battalion Operations Officer, my job was to essentially make sense of the deployment order and figure out how to support and resource the requirements and then make sure all training and equipping was done to standard and on time to execute the mission.

Again, the details remain a bit sketchy for me, but what we ended up with was a Task Force — Task Force Regulars — with Bravo Company (B/1-22) as the main effort, supported by the Scout Platoon, and various Headquarters and support personnel... and commanded by me — MAJ Reed — as the Task Force Commander. There were roughly 200 of us who deployed in April 2002 until June 2002.

JTF 160 was in support of Operation Enduring Freedom and most specifically the detainee operations at GTMO (Guantanamo Bay, Cuba). We were the only Army Infantry unit from Ft. Hood deployed at that time. Our unclassified mission was to provide external security for the detainee facilities, Camp X-Ray and then Camp Delta.

This included fixed site security, a quick reaction force capability, combat patrols, and when needed, augmentation to detainee transport. We also provided the security for the move from Camp X-Ray to Camp Delta when Delta was activated.

A few things from this mission jump out at me as some lessons that served me well later on. First, I worked for a Military Police (MP) Brigade Commander, with a staff of mainly MPs — active and reserve — and later a reserve Special Forces Brigadier General came in to run the entire show. I also needed to coordinate with the permanently stationed Marine Security Force — famously represented as Colonel Jessup's unit in a *Few Good Men* — that dealt with the active Cuban military threat. With a very small HQ staff, I did most of this coordination myself, which meant I needed to delegate — and trust tasks and responsibilities to my other leaders, in particular the B CO Commander, Jason W., and the Scout Platoon Leader, Chris M.

I was also working with units that I did not have a habitual relationship with and weren't even Army Infantry. This required patience, overcommunication, and creating shared understanding. Little did I know at the time that the skills I learned at GTMO would be extremely useful in less than a year when I deployed to Iraq.

Second, I needed to build a team. Although the units in TF Regulars were all organic to the same parent battalion, we were still disparate organizations who came together on the fly. Team building wasn't just going to happen with the sprinkle of pixie dust. Personal leadership matters. Physical training was huge; sporting competitions; rest and relaxation events... and a sense of humor... all worked to create a cohesive environment.

And we trained, maximizing the Marine ranges and training areas on GTMO. Trusting the officers and NCOs of the Task Force was critical but then holding people accountable when standards weren't met. Most important was soliciting feedback from my leaders on how to accomplish various tasks and missions. We were asked to do a lot with limited resources. We had to manage our "troops to tasks" to the gnat's ass. Leader input was critical.

Third (and definitely related to the point above) was organizing and training the unit for combat... or to meet the operational requirements of the mission.

Things happened quickly at Ft. Hood when we received the mission. I was able to fly to GTMO and do a Pre-deployment Site Survey (PDSS), talk to people on the ground, and meet the chain of command and staff. At the time, this was a very, very nascent mission in an immature theater, and in many ways, we were building the airplane in flight. But I gathered enough information to craft and resource our home station training and build flexibility into the organization, and I had the support of my leadership at Ft. Hood through the Division level.

Looking back on my time as the TF Regulars Commander, the lessons I learned absolutely impacted how I led — and fought — during the rest of my career. And this isn't the first time that I've said something like this. I may not have been the sharpest knife in the drawer, but I did realize that lessons carry over and I needed to pay attention to this.

Once the Army sorted out the long-term resourcing for the JTF 160 mission, we redeployed in late fall of 2002 — replaced by a National Guard formation. It was a good mission for us, but clearly not one that was best suited for an active-duty mechanized Infantry battalion. Upon returning home to Ft. Hood, we took some leave and prepared for a rotation to the National Training Center … while at the same time beginning to hear rumors about an invasion of Iraq and impending deployment. Cowboy Up!

The swirl around an invasion of Iraq was overwhelming. Shortly after returning from GTMO, I moved from my S3 (Operations Officer) position to be the Battalion Executive Officer (XO). My good friend and West Point classmate, Mike R., came to the Battalion to be Regular 3. I was now Regular 5. I immediately became a better man because of Mike. While truly a tremendous officer and leader, Mike was an even better human being. He made all of us better. Iron sharpens iron.

Shortly after Mike arrived and I moved to the XO gig, the battalion deployed to the National Training Center (NTC) for a full-scale force on force exercise, complete with company and platoon live fire events. It was at some point before, during, or immediately after this rotation that we found out we were deploying to Iraq as part of the initial ground invasion. It was just a confirmation of what we already suspected. More on that in a minute.

Back to NTC... a couple of memories. It was the last time I slept in a shelter half in my Army career. This archaic piece of kit was something I first used in my days at West Point. During the initial build-up prior to deploying to the "box" — aka, maneuver area — the battalion stayed in the cantonment area at NTC. This was basically a big-ass gravel parking lot with some overhangs to keep the rain off the shelter halves. I think the intent was to make you so miserable you couldn't wait to get into the box.

The second memory was that I literally didn't sleep as the Battalion XO — that is, until one of the Observer/Controllers "killed" me for eight hours and made me sleep. I falsely believed that I needed to go non-stop and that the battalion would fail without me. That was obviously not true. In fact, I probably caused more harm than good with little rest and downtime, and I learned a valuable lesson about "fighter management" and the impact of sleep and rest on combat readiness.

Upon our return from the National Training Center, the Battalion prepared to deploy for combat... and it happened very quickly. The initial plan was for the Division to invade Iraq from the north through Turkey, with the 3rd Infantry Division and the Marines invading from the south via Kuwait. At the Battalion level, we went through the orders process. The division published their operations order; the brigade did the same; and then we — 1-22 Infantry — issued our order. Simultaneously we went through the preparation of Soldiers and equipment for deployment, and some limited collective and individual training. Coming off a NTC rotation, we were at a high level of training proficiency and most of our key leaders were still in place.

Before the days of the Rapid Fielding Initiative (RFI) when new gear, gadgets, and equipment were literally issued to you days before — or at rampside — of getting on the plane, we were going to war with what we had on hand. Or, in the words of then-Secretary of Defense Donald Rumsfeld: "You go to war with the Army you have, not the Army you might want or wish to have at a later time." As I recall, he caught a lot of shit for that, but he was right.

We literally emptied the motor pool for the railhead... we took every piece of equipment — all Bradley Fighting Vehicles, Tanks, HMMWVs (none were armored), five-tons, LMTVs, etc... and we towed our broke-dick shit onto the railhead. We were going to war with what we had.

I'll never forget standing in our motor pool at Ft. Hood the day we deployed our equipment. We had just moved everything out of there and to the railhead. We would then load the vehicles on trains for movement to the ports and the boats for onward movement overseas. It was very surreal, made even crazier by the one loan Deuce-and-a-Half (2.5 ton truck) that sat off by itself in the corner of the motor pool. It was like a badge of honor... our last holdout giving me, the Battalion XO, the middle finger. It was supposed to go, but it wouldn't start. We were also out of towing assets to tow/drag this poor truck to the railhead.

To the rescue came then Specialist Mier, whose full name was Eliu Miersandoval. Born in Mexico and after moving to the United States, he enlisted in the Army after high school. He was a light wheeled mechanic who could literally fix anything—wheeled or tracked. He had a huge smile and an easy-going personality that made people want to be around him. Well respected, extremely competent, and a very hard worker, he was one of our best mechanics. As the Battalion Executive Officer and the officer with overall responsibility for the maintenance of the unit's vehicles, I spent a lot of time with our mechanics, and I was very much drawn to Mier.

So there we are—literally just SPC Mier and me in the motor pool—and this truck that won't start. Mier being Mier said something to the effect, "Don't worry, sir, I got this." While I wanted to believe him, I had my doubts. This vehicle had been on the deadline report for the past two years. In short, it was an old truck that the Army was phasing out.

That said, Mier's positive attitude and conviction in his abilities made me a believer, and sure enough, the vehicle turned over and SPC Mier drove that vehicle to the railhead with me in the passenger seat as the truck commander (TC). I remember very vividly laughing our asses off!

Sometime later, just days before deploying to Iraq, I also remember vividly promoting Mier from Specialist to Sergeant. I was honored to officiate his promotion. It was a beautiful spring morning in central Texas. We did the promotion just outside the battalion conference room in a large grassy area next to the headquarters. Mier was there with his wife and young son. Of course we were in BDUs, his clean and pressed for the big day, but slightly streaked and stained just a bit with the labors of his profession as a mechanic. It was perfect! No one was more deserving than him of this promotion. It was a testament to his exceptional performance and outstanding potential as a leader. I was honored and humbled to promote Mier from Specialist to Sergeant.

SGT Eliu Miersandoval (age 25) was Killed in Action on 31 January 2004 when his vehicle was hit by an Improvised Explosive Device near Kirkuk, Iraq. He was killed along with SGT Juan Carlos Cabral-Banuelos (age 27) and SGT Holly J. McGeogh (age 19). They were all mechanics on a logistical convoy enroute to Kirkuk. It was a parts run — picking up vehicle parts and other necessities to maintain and repair our vehicles. They were American Soldiers doing their job. They are heroes.

INVASION AND FOLLOW-ON

As with any deployment, there are always a few wildcards that will cause some anxiety. And as is well known and documented in the history books, Turkey wasn't playing ball. The 4th Infantry Division was supposed to attack into Iraq from the north while the 3rd Infantry Division and the Marines attacked from the south. That was the general scheme of maneuver anyway. So while all our vehicles and containers were on boats making their way to the Mediterranean Sea, we at home really had no idea when we would deploy, which caused great personal and familial stress. We also had no idea how long we would be gone once we left. No one did. The chemical threat was real, and we didn't have much confidence in our chemical protective gear and training — truthfully, the Army in general was not nearly as proficient as it should be in this task.

Finally, and most importantly, this was the real deal — people were going to die. Amidst this backdrop, we waited. At one point it even seemed possible we were going to miss the whole thing. Turkey said no. Our equipment was on a boat in the Mediterranean Sea, and the ground invasion had just started.

In March 2003, just as it appeared that we were not going to deploy, we did... and it happened fast. All our equipment on boats that was supposed to go to Turkey were diverted through the Suez Canal to Kuwait. As an aside, we sent Soldiers (volunteers) on those boats who were responsible for keeping an eye on our equipment, maintaining it as best they could, etc. Those poor guys ended up getting more than they bargained for with all the delays and rerouting!

The sense of urgency was driven by the start of the ground invasion and the "operational pause" that 3ID and the Marines ended up needing due to weather and the pace of the fight... essentially, they were outrunning their supply lines. That was my read of it anyway.

My much larger concern as the Battalion Executive Officer was to deploy as quickly as possible with the TORCH Party — the advance party to the advance party (ADVON), which is in advance of the main body — TORCH then AD-VON then Main Body. Basically, I would deploy with a small number of officers and NCOs with the right skill sets — primarily logisticians and decision makers — to set the conditions for the follow-on advance party of drivers and mechanics.

All of this was intended to get our living space and motor pools set up and to begin the offloading of our equipment from the boats. The main body of the battalion would soon follow.

We moved quickly. Saying goodbye to my family was hard... it always is, but because it happened so fast a lot of the anxiety that comes with the anticipation of a deployment didn't exist. Don't get me wrong. It still sucked. Unlike later deployments, however, for this one we had no idea how long we were going to be gone, nor did we have a sense for the type of fight we were getting into... which was a good and bad thing. And taking your "death" photo is always disconcerting.

This is the photo that would go to the media in case you were killed or wounded in action. It was a not-so-subtle reminder of the realities of combat.

In early March, I deployed with a small contingent from the Battalion and other representatives from the Brigade and across the Division. We went to Camp Pennsylvania in Kuwait. We arrived shortly after Sergeant (SGT) Hasan K. Akbar of the 101st Airborne (Air Assault) Division murdered two of his fellow Soldiers and wounded others in a horrific tragedy for which Akbar is imprisoned and on death row. While this felt like a bad omen for us, our immediate issue was the multiple SCUD attacks/warnings that kept us on edge and wondering what might lay beyond the berm when we began the invasion.

With the arrival of the main body — the battalion, the brigade, and the rest of the division — the priority became offloading the boats and organizing for combat. Reflecting back on all of this, we — the leadership — probably sacrificed safety for speed, but the sense of urgency was clear. The 4th Infantry Division needed to get into the fight. What I would characterize as a follow and support mission, the Task Force Ironhorse mission was clear — offload the boats, get our vehicles and containers to Camp Pennsylvania, fuel, arm, and prepare and organize for combat. And to do all of this at the speed of war.

As the XO, I was going to be in the trail of the battalion with most of our support vehicles. I was fortunate enough to have my own Bradley Fighting Vehicle (BFV), and I also had my High Mobility Multipurpose Wheeled Vehicle, or HM-MWV. I decided to cross the berm in my HMMWV. I'm not really sure why, as the Bradley afforded more firepower and protection, but that was the plan. The plan was also to recover any vehicles that broke down along the route and tow them north with us into Iraq.

That plan shit the bed in the Camp Pennsylvania motor pool when we couldn't even get some of the vehicles out of there. In what I have characterized as the Trail of Tears, we left our broken-down vehicles where they went down, and cross-leveled people and equipment into other vehicles as we moved north.

Task Force Ironhorse invaded Iraq on 14 April 2003. I personally crossed from Kuwait into Iraq a few days later and joined the fight... to use that term loosely. 1-22's destination was Tikrit to conduct a relief in place with the Marines. As I bounced around in my HMMWV over the course of the hundreds of miles to Baghdad, where we were to link up (as the trail party) with the rest of the battalion, I just remember trying to stay awake, getting lost, dealing with broken vehicles, confounding logistical resupply points, unreliable communications, and crossing through multiple friendly units that were manning checkpoints... but fortunately no enemy contact — otherwise that would have been a total shitshow.

I also don't recall even having a map. It was amazing that we reached Baghdad and even more amazing that we linked up with the rest of the battalion. I would summarize the "attack" north as a safety officer's worst nightmare and less of a combat action.

Things changed when we were preparing to leave Baghdad, however. After a quick refit on the outskirts of Baghdad and some internal Battalion re-organization, Task Force Regulars prepared to move north to Tikrit. We were heading to Saddam Hussein's hometown to conduct a relief in place with the Marine unit currently there.

Again, as the Task Force Executive Officer, I was in the trail of the task force movement order of march with the logistical element. Given that we had to secure ourselves — i.e., there was no dedicated combat power, like an Infantry platoon, to provide security for the soft skin and lightly armored vehicles — I decided to command and control this effort from my Bradley. With the 25mm chain gun and 7.62 coax, it was extra fire power to secure the rear element, and I was offered the protection of an armored vehicle.

While I didn't know what to expect, I also didn't really think we'd be in a gun fight. Most of the route had been cleared and was secured with patrols and checkpoints of the forward land-owning units.

However, right out of the gate, and I mean literally out of the gate, we took contact. The formation assembled in a secure and guarded area. As we rolled out of the perimeter security, we immediately took fire from a building directly to our front. This was my official "first contact" and it was underwhelming. It went something along the lines of hearing the deflection of rounds off my Bradley and wondering what the noise was. It didn't register that we were taking contact. In fact, I remember asking my driver and gunner, "what was that?" To which they responded that they thought we were taking contact and that I ought to drop down in the commander's hatch.

At the time, like a good Bradley Commander, I was at name-tag defilade in the turret. My gunner scanned the area and saw that the fire was coming from across an open field in what was an abandoned, semi-bombed out building. To his credit — and very much to his training — he immediately dialed up the 7.62 coax and smoked the location where he saw the fire originated. Problem solved. He didn't wait for me to say anything, which is good, because I was still trying to figure out what was going on. With that, we headed north.

It didn't take terribly long to get up to Tikrit and I don't recall a whole lot of handovers with the Marines... it kind of felt like a high five... and with that we took over the area of operations. The Battalion Task Force's area of operations extended from Bayji in the north down to Tikrit and then points east and west. The Task Force Ironhorse Headquarters set up in our footprint — the Division Main, or D-Main — which brought with it security requirements and other miscellaneous responsibilities for us.

The immediate task at hand was getting a lay of the land, understanding the environment and the enemy, as well as a reorganization of our assets to accomplish the mission — a mission, which at the time, was a bit unclear. It was a strange time in the war.

While Saddam's Army was defeated, the nascent insurgency in the form of the Fedayeen was proving to be an irritant and quickly developing into a legitimate threat. The local population was caught somewhere between jubilation and confusion and began looting and stealing everything from pots and pans to chain link fencing. If it wasn't locked down, it was getting looted. Our mandate to prevent this wasn't clear. The fundamental issue was that there was no governance and no services — water, electricity, security — and the Iraqis were looking to us for answers.

So, we figured it out... and I mean, we literally figured it out. In my view, this is where "by, with, and through" Iraqi counterparts was born. We couldn't do this without the support of the Iraqis.

We were the only security at the time, but that was the easy part. We knew how to do security. The hard part was sorting out the politics and tribal affiliations — who were the key players; what were our authorizations in terms of dealing with them; drinking a ton of tea to establish relationships; figuring out who spoke some English so we could use them as interpreters; talking to people as best we could and building trust.

The wild card in Tikrit was that this was Saddam's hometown (actually it was al-Ouja, just on the outskirts), so, in general, the populace was not overly helpful. No one taught us or trained us on how to do all of this, but we managed to put together a plan on what made sense and drew on our previous experiences in peacekeeping and stability operations (Haiti, Somali, Bosnia, etc.). Slowly but surely, we started to get some semblance of governance in order, turned on the water, and began establishing security.

But we were still at war — or at least in combat — which meant that we were still fighting an enemy who very much hated us, supported Saddam Hussein — who was still on the run — and by extension, hated those Iraqis who were supportive of the United States and the coalition. Given that context, Task Force Regulars and the Raider Brigade (1st Brigade, 4th Infantry Division) suffered their first casualty of the war on 25 April 2003 when 1LT Osbaldo Orozco (age 26) was Killed in Action. He was a platoon leader in Charlie Company.

Given the nascent combat theater at the time, communication with families back home was Desert Storm-like correspondence, and notification to the rear detachment was slow, clumsy, and disjointed. Messages were mixed and confused. As an Army, we got better at this over time, but, regardless, Osbaldo's death was a tough reminder for us all that the stakes were high. We bungled the notification back to the rear detachment and there was confusion on what exactly happened, or who was even killed. Eventually that all got sorted out, but the reality was that a Soldier was dead and sadly he would not be the last.

With a platoon now hurting and needing leadership, the next man up was 1LT Jason L. who was working in the S3 shop as a staff officer. The young officer was fairly new to the unit and had not been down to a platoon yet. He immediately stepped into the platoon to provide leadership for the 30+ Infantrymen who were in pain, needing guidance, and still had a mission to accomplish. Jason was the next man up and set the standard for the many others who would need to do the same throughout the Global War on Terror.

As April transitioned into May, and May into June, the pace of operations picked up — both on the combat and non-combat side. At some point in here the Vice Chief of the Staff of the United States Army, General Jack Keane, visited our battalion. He told us we'd be here a long time. That was it. While I think he really meant the Army in Iraq (and he was right), it was the first time someone gave us an indication of how long we — our unit — would be deployed.

In short, it was until the Army told us to go home (which wouldn't happen until April 2004 for 4ID). For many, it was a kick to the gut — disappointing that we had no clarity on when we would go home. On one hand, the unknown added to the stress and anxiety. On the other hand, it made it clear to focus on the mission at hand. While it may sound naive, I dialed in on what I could control and let the decision makers figure out when units would deploy and redeploy.

If I had to pick one moment when I think everything changed, it would be 7 June 2003. Up until that point, aside from the death of LT Orozco, it felt like TF 1-22 had things under control in our area of operations. In a brazen and coordinated effort, insurgents attacked the Civil Military Operations Center (CMOC) in downtown Tikrit in broad daylight. The CMOC was our first attempt at community outreach. Locals could come to the CMOC and file claims, ask questions, etc. It was jointly manned — U.S. forces and Iraqi government officials — but it was 100% guarded by U.S. forces.

And on 7 June 2003, the Fedayeen attacked. In what turned out to be a well-coordinated insurgent effort, the attack was repelled by our local security and the Quick Reaction Force, but not before PFC Jesse Halling (age 19) was killed, Jesse was an MP whose unit was attached to the Task Force. Jesse was manning his fighting position, engaging the enemy, on the day he died.

As the Battalion Executive Officer, I was in the Task Force Tactical Operations Center (TOC) coordinating the Casualty Evacuation (CASEVAC) and working the combat support — fires, aviation, Quick Reaction Force (QRF). The QRF was the Task Force Scouts led by 1LT Chris M., an incredible officer who I would serve with at various times in my career. He was a warfighter. Moving at the speed of combat, Chris and his platoon responded to the contact at the CMOC, secured the area, and rounded up several persons of interest who were processed to the Division Holding Area for further questioning.

Coordinating all of this from the TOC was a chore, in addition to keeping our higher headquarters informed of what was going on. My most significant lessons learned from that combat action were the importance of calm communication and tactical patience, allowing the situation to develop, staying in control and being unemotional on the net, and timely and decisive decisions. This has an impact up and down the chain of command and instills confidence in all those involved in the situation. It is the art of leadership in combat.

THE RAIDER BRIGADE

"Assholes will not dictate terms to me."

— unattributed

In late summer of 2003, I moved from my Battalion XO job to be the Brigade Operations Officer (S3) for 1st Brigade, 4th Infantry Division... the Raider Brigade. This was a pre-planned move that required the intervention of the Division Commander, MG Odierno. It wasn't common for a Major to serve in three Field Grade positions — in my case, Battalion S3, Battalion XO, and now Brigade S3. My well-intentioned branch/personnel manager wanted to move me to another assignment out of 4ID after my time was complete as the XO. A phone call from the CG changed that. I'm forever indebted to then-MG Ray Odierno for the opportunity. The timing of the move was delayed/in flux dependent on battalion and brigade changes of command, combat operations, etc. But here I was... Raider 3. I made the physical move from the 1-22 IN footprint to the Raider Command Post (CP), which was on the southern end of Tikrit.

I inherited an incredible group of core staff Captains — Geoff, Barry, Mario, Mike, Mark Paine, and Ian Weikel. These were hard working Captains who were doing their time on the Brigade Staff prior to taking company command. Being in a combat zone as a Brigade Staff Officer instead of a company commander can be tough, and maybe even a little depressing. You want to be in the fight...period. But every one of these officers were consummate professionals and recognized that they had a job to do.

I learned a lot from them. Their jobs may not have been sexy or elegant, and it was often thankless, but they are the ones who wrote the operations plans, manned the radios, submitted the reports to the Division, kept the Battalions informed, monitored the current fight, and were responsive to the commander. They did it with pride and professionalism, and humor... a lot of humor!

A couple of years later, on a subsequent deployment to Iraq, Ian Weikel was killed in action as a Troop Commander in April 2006; Mark Paine was killed in action as a Company Commander in October 2006. These West Point classmates (USMA 1997) and close friends are together in eternity, and they are buried next to each other at Arlington National Cemetery. Ian and Mark... "two monkeys banging on a keyboard." I love those guys. I miss them.

My boss, Raider 6, COL Jim H., was a tremendous warfighter. I had a front-row seat to battlefield genius while at the same time earning a Ph.D. in how to command troops in combat. When I took over as Raider 3, the insurgency was at full throttle and Saddam Hussein was still on the loose. Theories abound as to his location, and the endless "sightings" and rumors could easily derail the brigade if we weren't careful.

The cooperation with the special operations task force was textbook "hand in glove." There was very little white space between us. Raider 6 fostered that. He demanded that we share information with our special operator teammates and provide them whatever they needed, whether it be a quick reaction force or an additional platoon to provide an inner cordon on an operation. It was this kind of cooperation that eventually led to Operation Red Dawn and the capture of Saddam Hussein.

COL H. would often say that "the war is not won on the FOB [Forward Operating Base]." If there was a gunfight in the Brigade's Area of Operations, he was there. And so was I as his S3. No one patrolled more than his command patrol. I was part of that patrol, either in the lead with my vehicle, or second in the order of march. Our command patrol rolled with three thin skin HMMWVs. Raider 6 was in one; I had the other; and we had a gun truck. We were like any other combat patrol in the Brigade Combat Team. "RPG Alley" became our second home and where I personally employed *Battle Drill 2: React to Contact*, on multiple occasions. Fortunately, our command patrol was good at it!

This was the only way Raider 6 was going to have a full appreciation of the enemy, the terrain, and the environment in which his Soldiers were fighting. He was aggressive in combat, and he drove the unit hard. He expected much of us, but he expected more of himself. He exacted excellence from the Soldiers and leaders in the unit through his personal leadership, personal commitment to excellence, and the standards that he demanded. He listened to their input, he cared about their feedback, and he was transparent in how he communicated.

It was his personal combat intuition, sense of the enemy and the terrain, and his ability to bring together a combined arms team — a diverse and talented team — that led to the capture of Saddam Hussein. It was always "mission first." COL H. accepted, understood, and embraced this. He lived and modeled this. It was about what was best for the unit and accomplishing the mission.

Raider 6 is the reason Saddam Hussein was captured on a dark, cold night in Iraq on 13 December 2003. He was the senior commander on the ground. He was the commander of the formation that conducted the operation — a formation that included all elements of Army combat power... Infantry, armor, artillery, aviation, scouts, engineers, logistics, military intelligence, and special operations forces. He worked closely with the Special Operations Task Force to build an intelligence picture and execute an operation that was textbook.

Much of the hard work, however, was done long before Operation Red Dawn. As the Raider Brigade Operations Officer, I had a front row seat for arguably one of the finest intelligence and operations workups ever. The Brigade put together a network of Saddam's associates, family members, and colleagues, essentially from scratch. The painstaking work grew out of targeted operations that pieced this network together, driven by Raider 6's untiring battlespace presence.

Done in conjunction with our special operations teammates, it was a textbook example of conventional-special operations partnership and cooperation. It was also a great example of the complete disregard for who gets credit, and instead working together to get the mission accomplished, and in this case, it was the kill/capture of Saddam Hussein.

Operation Red Dawn was an operation like any other, just like the many others we had executed on multiple occasions in Iraq. We had fairly reliable intelligence about the location of Saddam. Orders were issued over the radio and via digital systems. The raid force assembled in an abandoned granary north of Ad Dawr for a quick leader's huddle and final coordination. The Special Operations Task Force was the main effort and would conduct the actions on the objective to either kill or capture Saddam Hussein. The Brigade Reconnaissance Troop, commanded by Captain Dez Bailey, would set the inner cordon.

In many ways, Dez and his Troop were the action arm of the Brigade Combat Team. Beautifully articulated in his book, *Recon 701*, Dez commanded a company that was defined by its audacity, innovativeness, and agility. The Troop were the habitual and preferred partners of the Special Operations Task Force in all SOF operations in the Brigade's battlespace — they moved at the speed of war.

Like any other night, the Troop set the inner cordon and ensured the objective was locked down to perfection. Led by a warrior and defined by a warrior culture, Operation Red Dawn would not have been successful that night without the personal leadership of Recon 6, CPT Dez Bailey. Desmond Bailey died on 19 July 2023.

OPERATION RED DAWN

"We got him." With those words, Ambassador Paul Bremer announced to the world that coalition forces had captured Saddam Hussein. This marked the culmination of months of intensive intelligence and operational work. The raid itself to capture Hussein was nothing spectacular. It was like any other of the hundreds of raids we'd executed since we deployed to Iraq in March 2003.

When the sun came up on 13 December 2003, there was nothing unique about that day. However, that evening, at approximately 8:26 local time, the enemy was caught off guard by an aggressive, determined and agile combined force. The signature event of Operation Iraqi Freedom up until that point, Operation Red Dawn, was textbook with all forces executing flawlessly and swiftly.

The capture of Saddam Hussein was a perfect example of Conventional Forces and Special Operations Forces (SOF) cooperation. It was hand-in-glove — a team effort between the 4th Infantry Division Headquarters, Special Operations Forces, and the 1st Brigade Combat Team. The coordination between units was the best that I was personally a part of at any point in my career. No one cared about who received the credit. It was Warrior Professionalism at its finest. All units understood that each needed the other to piece together this puzzle and leverage the resources and expertise within our respective organizations.

Since its arrival in Iraq, the Raider Brigade had been developing intelligence designed to find not only the Top 55 most wanted individuals — the Department of Defense published list of the Top 55 most wanted members of the Saddam Hussein regime — but to locate and kill or capture those lower tier operators and facilitators who were providing the funding, coordinating opposition, and encouraging operations against coalition forces.

Each day another piece of the puzzle fell into place, which led to more of the key players being identified and located. The Brigade began to build highly visual "link diagrams" that showed the structure of Saddam Hussein's personal security apparatus and the relationships among the persons identified. These link (or network) diagrams showed everyone related to Hussein by blood or tribe. These family diagrams led us to target lower level, but nonetheless highly trusted, relatives and clan members potentially harboring Hussein and helping him move around the countryside. The circle of bodyguards and mid-level military officers, drivers and gardeners protecting Hussein was often likened to a Mafia-type organization. It was built on hierarchy and who had access to Saddam himself.

In early July, the Brigade, having operated in the Tikrit area since April 2003, had a mountain of information in its database pertaining to Saddam Hussein, family members, close associates, and unrelated insurgents. It was a database that would increase ten-fold prior to the capture of Saddam. As the amount of information and intelligence increased, so did the analysis.

From this analysis, four names kept coming up and

seemed to be linked. The Brigade's Intelligence Section, in coordination with our Special Operations teammates, took those names and compared them to what we already had and started making links out of them. In less than a week, those four names had grown to more than a hundred. In a month, the names and families filled a 36-inch by 36-inch sheet of paper. Over time, we began to understand how each name was connected to the next.

This included assigning roles and positions to certain people within the network — for example, chief of staff, chief of operations, personal secretary, etc. These were not necessarily positions the individuals occupied prior to the fall of Hussein, but instead were based on our understanding of the role they were likely filling in support of the insurgency or Saddam's underground operations. We assigned these roles from our assessments of various personalities and current intelligence and information reports. Such a process helped us focus our efforts in determining those who were closest to Hussein and their importance.

Over the days and months, our unit continued to track how the enemy operated. We tracked his trends and patterns, examined the tactics the enemy employed, and started to connect the enemy tendencies with the names and groups on our tracking charts. Our Intelligence team made minor adjustments to the template and kept looking at all the critical data points to find what we may have missed.

It was the capture of two key associates of Saddam Hussein on 7 November 2003 that confirmed for us that our template was accurate and that by continuing with the current operational focus, we would eventually succeed in capturing the ousted dictator. Toward the end of November, after several weeks of little to no new information relevant to Saddam and his network, a series of operational events led to an abundance of information and new intelligence about the resistance and the whereabouts of Hussein.

The result of this latest intelligence was a series of raids, all designed to capture key individuals and /or leaders of the former regime that could eventually lead us to Saddam. Each raid resulted in more information that led to the next raid. This cycle continued as several mid-level leaders of the former regime were caught eventually leading into the inner circle of those most trusted by Saddam.

On the morning of 13 December 2003, I was at a Division Rehearsal at the DMAIN for an impending operation in vicinity of the town of Samarra. Samarra was not in the Raider Brigade Area of Operations, so we were a supporting effort to this Division-level attack. At some point during the rehearsal, I received a message that Raider 6 was trying to reach me. I stepped out of the room and found a workable phone to call back to our TOC (Tactical Operations Center). I got on the line with the boss, and he instructed me to get down to the Task Force's (TF) command post immediately.

A source was captured in Baghdad the night prior and he might be the missing piece that could lead us to Saddam. SOF was flying him up to Tikrit in the next hour, and Raider 6 wanted me to link up with the SOF Team Leader and see if this individual was the one we had been looking for — the missing link.

By the time I arrived at the TF command post, the source was already there. After some discussion, we confirmed that he was the messenger we had on our link diagram who could take us to Saddam. I immediately called Raider 6 and let him know. Within short order, he was on his way to meet us at the TF location. Upon Raider 6's arrival, in conjunction with the SOF Team Leader and his key leaders, we drew up the tactical plan for the operation that later became Red Dawn. I drew the scheme of maneuver on butcher paper, and I had my planners back at the TOC transcribe it into an Operations Order format with graphics and control measures. We shared all of this over the radio and FBCB2 with the participating units.

The plan was to go heavy — maybe heavier than we needed. Several months prior to this, Saddam's sons, Qusay and Uday, holed up in a building in Mosul. An extended gun fight took place before they were eventually killed. Raider 6 didn't want an extended gun fight. At any sign of resistance, we would use the appropriate force necessary, within the Rules of Engagement, to eliminate the threat. To this end, we had a troop from the Division Cavalry Squadron in reserve. Also in support of the operation was our organic Engineer Battalion, 299th Engineers, who had a unit block any egress to the west along the Tigris River. We employed the Division's aviation element to screen and provide supporting fires if needed.

The Brigade's attached field artillery battalion, 4-42 FA, owned the battlespace that encompassed Ad Dawr. They also provided fire support to the brigade. Their daily mission set was tremendous, and they did a great job. For Operation Red Dawn, they provided the outer cordon, locking down the key points around the objective.

As an aside, the Battalion Commander was on his well-deserved R&R leave, so the Battalion Executive Officer was in command of the battalion at the time. It proves the point that good units are not dependent on a single person. The "next man up" was not a trite phrase. Good units groomed leaders to step up and take charge when necessary, and 4-42 was a good unit.

The Brigade Reconnaissance Troop (BRT) under the command of Dez Bailey had the inner cordon. As mentioned earlier, Dez and his Troop had a habitual relationship with the Special Operations Forces in our AOR, or Area of Responsibility. There was no question that they would lock down the immediate objective area with the Task Force as the assault force and the main effort. The Brigade TAC, aka the Raider Brigade Assault Team, served as the command-and-control (C2) element with Raider 6 as the senior officer on the ground with overall command of the formation.

The Raider TAC, the BRT, and Task Force elements linked up at a granary north of Ad Dawr, just on the east side of the bridge that spanned the Tigris River from Tirkit to the west. After this quick leader's huddle, at approximately 6:00 PM, under the cover of darkness and with lightning speed, the Raider Brigade forces were positioned and began movement toward the objectives north of Ad Dawr. Time and surprise were vital. Experience taught us that rapid exploitation of information about the enemy was important. Any window of opportunity between the moment the information was received and the time that the operation was executed could be just enough for the target of the raid to escape.

At approximately 7:45 that same evening, a coincidental, but extremely fortuitous power outage resulted in all the lights going out in Ad Dawr and the surrounding area. With moonrise not until later, the city was completely dark, which aided in concealing the combined force's movement as we entered the city. There was no civilian traffic along the main road into town. The force entered the objective area at 7:55 and had forces on Objectives Wolverine 1 and Wolverine 2 by 8:00 PM. There was nothing significant to report on either objective.

Despite an earlier reconnaissance, the source (who was brought along on the operation to positively identify Saddam Hussine) indicated we were in the wrong place. Raider 6, in coordination with the Task Force leadership, expanded the search further to the northwest of Wolverine 2 based on what could be discerned from the source's information.

The source led the assault team to a mud hut. One of the occupants at the location attempted to flee, but he was detained. Another occupant vehemently denied knowing anything about Saddam Hussein and attempted to lead the assault team away from the objective. However, before he could do so, the source identified and pointed out the location of a potential hiding position near the edge of an orchard.

The area was a small walled compound with a metal lean-to structure, and a mud hut. During the search, a spider hole was detected. The spider hole's entrance was camouflaged with rocks and dirt. The assault team cleared the entrance to the hole of dirt, debris, and mats, and removed a Styrofoam insert.

The person inside of the hole put his hands in the air and responded to the question, "Who are you?" with "I am Saddam Hussein, the President of Iraq, and I am willing to negotiate." With that, a Soldier from the assault team responded with, "President Bush sends his regards," and Saddam Hussein was pulled from the hole. After a search of the area, a helicopter landed and evacuated Saddam Hussein. By 8:15 PM, initial reports of "JACKPOT" were sent to the Brigade Commander. At 8:26 "JACKPOT" was confirmed.

I was personally located on the outside of the walled compound near my HMMWV. As the Brigade Operations Officer, my primary role at that point was to keep our higher headquarters informed and to work the coordination for any additional combat power or put into motion Raider 6's orders and instructions. When the call came over the Brigade's internal radio net that we had JACKPOT, I made the call to the Division Main and notified them of JACKPOT.

Saddam Hussein had approximately $750,000 dollars in U.S currency in a green metal container on his person when captured. The Raider TAC secured that money and brought it to the Division Headquarters and presented it to General Odierno later that evening. We then turned the money over to the team responsible for conducting the Sensitive Site Exploitation (SSE) of the objective area. The money and other items collected from the objective were catalogued and forwarded to the appropriate authorities for further evaluation and processing.

Before the Raider TAC left the objective area for the Division Headquarters, we put the Styrofoam insert from the spider hole into the back of my HMMWV. I'm not sure who did that, but for whatever reason, there it was. When we returned to FOB Raider later that evening, I put the insert in the room I was sharing with Raider 5, Troy S., the Brigade Executive Officer, and Raider 7, Larry W., the Brigade Command Sergeant Major.

And there it sat, about 5 feet from my cot, for several weeks. Honestly, I forgot about it. The operational tempo of the fight didn't slow down, so I had other things to worry about it. One day, Troy asked me if I knew where the insert was located. Division was looking for it. I told him in our room...5 feet from my cot! Troy also forgot about it. That was the last time I ever saw the insert. From my understanding, after turning it over to Division, the Styrofoam insert made its way back to the museum at Ft. Hood, Texas.

After Saddam was captured, there was an immediate shutdown of all external communications — email, social media, phones, etc. The good news was that this was 2003, and the communication infrastructure in the theater was still immature, so it didn't take much to shut this down. The announcement on the capture was going to come from Ambassador Bremer in Baghdad the next day on 14 December. I sent CPT Mike W. to Baghdad to put together the briefing for the announcement. Mike was one of the Brigade planners.

At the time of the announcement, all the Brigade Staff was gathered in the TOC, a large auditorium type room in one of Saddam's palaces, with multiple TVs, communication systems, planning areas, and the operations command center. While there was a lot of activity in the room, it was really very simple — only what was necessary, pristine, and professional. It was not a place for social gatherings. But for today's announcement, it turned into something of an old-school company dayroom.

I was standing next to our primary planner, CPT Geoff M., waiting for the Ambassador's press conference to begin. Geoff was the person responsible for turning Raider 6's guidance and intent into an operations order. He and I worked very closely making sure we captured things correctly before we pushed out an order to the Brigade units.

During the planning for Operation Red Dawn, I was co-located with Raider 6 and most of our time prior to the execution was spent coordinating with the SOF unit and visiting the Brigade units involved in the operation. Raider 6 wanted to ensure that there was no white space between his guidance and interpretation — hence the unit visits and back briefs. As a result, I transmitted the tasks, purpose, guidance, and intent over the radio to Geoff.

Geoff produced all graphics, control measures, and the physical order based on what I shared with him. He then put together the Division required contingency operation presentation and forwarded this to the Division HQ. This was a PowerPoint packet that included the major muscle movements — task organization, intel update, mission, and concept of the operation. At this point in Operation Iraqi Freedom, there was no mandated format for naming operations. We were basically making up names — names that made sense — for our operations. In short, I had no idea what we had named the operation to capture or kill Saddam Hussein.

In the planning process, our SOF teammates labeled the objectives "Wolverine" based on their internal naming conventions. Standing next to Geoff in the TOC waiting on the announcement, almost as an aside, I asked him, "By the way, what did you name this thing?" Geoff responded, "Red Dawn." I told Geoff that was a great name especially since the objectives were named Wolverine 1 and Wolverine 2. Geoff looked at me and said something like, "Sir, I have no idea what you're talking about." I said to Geoff, "You know, Red Dawn, the movie...Wolverines!"

Geoff still wasn't tracking. I then asked why he named it Red Dawn. Geoff said, very matter of factly, "I always wanted to name something Red Dawn." He may have also said something about it being a cool name. So, when you look at the cover slide of the media brief, which contains a helicopter, framed in a red dawn... now you know. Once the announcement was made and the operational name was attributed to the movie, I didn't bother trying to get it back in the box. That said, the operation was not named for the movie.

Now that the announcement was made, we had to prepare for the press conference that was going to occur the next
day at the location of the capture. This required the same level
of planning and preparation as any major combat operation
we had performed to that point. The insurgency was in full
throttle, and we saw the threat to the press conference as very
high.

It would be a major score for the insurgents to disrupt
the event, especially given the number of reporters and media
outlets. Security had to be airtight. We had to lock it down,
while maintaining operations throughout the rest of the Brigade's battlespace. It went off without a hitch.

Naturally, our SOF teammates were not going to be at
the press conference. One thing that was not clear was what
exactly Saddam said when he was captured. Only the SOF assault team knew the answer to this. There was no doubt that
this would be a question asked at the press conference.

On the evening of 14 December, as we were going
through the final rehearsal for the press conference the next
day, I tried to call over to the SOF Command Post. Unfortunately, the phone lines were down, which meant I needed to
get my patrol together and drive to the opposite end of Tikrit
to the SOF CP to ask the operators. Every patrol is a combat
patrol, and fortunately we arrived and returned without inci-
dent.

The following day, at the press conference, Raider 6 and
Raider 7 were handling the media, providing an overview
of the operation, and taking questions from the reporters. I

was around the corner out of the line of sight, behind a HM-MWV, talking with Dez Bailey and Geoff. I heard Raider 6 call my name. I threw the dip of Copenhagen in my mouth onto the ground and walked around the HMMWV to face the throngs of media.

Raider 6 asked me what Saddam said when he was captured. I recall finding this a little odd because we had prepped the boss for this question, but I guess he had forgotten. So, I said, "I am Saddam Hussein, the President of Iraq, and I am willing to negotiate;" and a Soldier responded with, "President Bush sends his regards." The first time I recounted all of this to the media I wasn't very loud. Raider 7 tightened me up and told me speak up! And I repeated it.

Once the press conference was complete, the Brigade ended up securing the objective area for several more weeks until the nascent Iraqi Army was able to take over this responsibility. The requirement to provide security was taking away much needed combat power from other operations, so we were glad to hand this mission over to the Iraqis.

I naively believed that the capture of Saddam Hussein might have been a turning point for us in the war. Unfortunately, as history has documented, that was not the case. The intensity of the war picked up and the capture of Saddam faded into the background for us. The operational tempo kept us focused on the task at hand.

BACK TO GRADUATE SCHOOL

When the Division redeployed from Iraq in late spring of 2004, I didn't stick around long. I was hired for a rotating Ph.D. position back at West Point in the Department of Behavioral Sciences and Leadership. I never really thought about the long-term plan other than that I would go and teach at USMA after battalion command (assuming I was selected for command). Admittedly, I thought the odds were pretty good that I would be selected. I had three Key and Developmental jobs as a Major (BN S3, BN XO, and BDE S3) during 3+ years, and I did well. I also had two combat/operational deployments during that time.

So, the plan was to go back to the University of Maryland where I earned my master's degree in Sociology and completed my Ph.D. While I was not quite ABD (All But Dissertation), I had completed most of the coursework required for my Ph.D. during my master's program. As an Army fully funded graduate student, I was required to take a certain number of credits. During my Master's work, my advisor, Dr. David Segal, recommended I take courses that would count towards my Ph.D. I never had any intention of returning to graduate school, but I was in the Army, and I did what I was told! It turned out to be excellent advice. Thank you, Dr. Segal!

My timeline was tight. I entered Maryland in the summer of 2004, and I had two years to earn my Ph.D. I was selected for Battalion Command with a change of command date in the summer of 2006. To say that the two years at Maryland were stressful would be an understatement. In addition to my own guilt and inner turmoil about being in graduate school while the wars in Iraq and Afghanistan raged, I had a lot of work to complete.

I retook some courses, and I took a few new ones to prepare myself for my specialty examinations in Social Psychology and Military Sociology. The bigger issue at hand was writing and defending my dissertation. This is a process that takes people several years to complete, and I needed to get this done in 18 months. Fortunately, David Segal was an incredible advisor and mentor. At his suggestion, I ended up writing my dissertation on social network theory and using that theory to piece together roles, relationships, and networks to better understand an insurgency.

I used the network of Saddam Hussein that we built for Operation Red Dawn and declassified the data. Given that I was on the Army dime and earning an academic degree, this was a win-win.

There is no doubt that studying Sociology made me a better officer and leader. Soldier and Scholar are not mutually exclusive and I would argue that our Army needs more Soldier-Scholars. The study of Sociology pushed me to think and reflect in ways that helped me see another perspective, engage with people and cultures not like me, and structure arguments and discussions that were unemotional and rational. The classical and post-modern Sociological theorists that I studied taught me how to reflect and converse in an increasingly complicated world.

My dissertation, *Formalizing the Informal: A Network Analysis of an Insurgency*, contributed to both the Army and the discipline of Sociology, and I was awarded a Doctor of Philosophy in June 2006. There wasn't much time to celebrate, as I needed to get to Fairbanks, Alaska to take command of 2nd Battalion, 1st Infantry in the 172nd Stryker Brigade Combat Team in August 2006.

5. BATTALION COMMAND

THE PERFECT INTERSECTION

A mentor once told me that battalion command is the perfect intersection of experience and exposure. It's the first time in your career where you are nearly the most experienced Soldier in the unit and you are the "on the ground" leader... literally, in the fight. I also saw it as being similar to the veteran quarterback of the football team — you've been in the league for a few years, at the line of scrimmage, making the calls, changing the plays, making decisions, making sure the offense is drilled and rehearsed, holding teammates accountable; but also getting hit, dirty, and in the game.

Needless to say, I was thrilled to take battalion command, especially at the height of the Global War on Terror. To have the opportunity to command, and then to command that unit in combat, was a tremendous honor, and responsibility. Battalion Command selection was different back then. If you were eligible for command, you could select what category you wanted to compete in — the options were something like tactical, installation/garrison, and training. To increase one's odds for selection, our branch managers advised us to compete in all three categories.

Perhaps arrogantly, I competed only in the tactical category. I felt like I had a strong file, and frankly, all I wanted to do at this point was command an Infantry battalion in combat. I also wanted to go to Alaska... fortunately, this was something that I was able to request given the command selection process at the time.

And it all worked out... I was truly blessed. I was selected for tactical battalion command, and I would assume command of 2nd Battalion, 1st Infantry Regiment, 172nd Stryker Brigade Combat Team (SBCT). Stryker formations were still relatively new at that point in my career. While I'll never know for sure, I believe my background of light and mechanized infantry experience made me a good fit.

Lieutenant General James Dubik spearheaded the creation of the Army's first Stryker Brigade Combat Team and developed the training methodologies for these formations. He is also credited with the following quote that served as my north star as a Stryker Commander: "The Army envisioned a force that could fight like Rangers, think like Special Forces, and possess the mobility of a heavy force."

I initially wasn't sure if the 172nd SBCT was in Fairbanks or Anchorage (it didn't really matter to me) ... but it was Fairbanks, which would turn out to be life changing! I was set to take command in August 2006. The unit was currently deployed to Iraq and was finishing up a 12-month tour. I left Maryland (where I was living at the time) in early August for the drive to the Last Frontier. I was in contact with the current battalion commander and several other leaders who had already redeployed from Iraq back to Alaska with the advance party. To say that I was fired up would be an understatement.

I drove our Toyota Sequoia loaded up with as much as I could fit in the car, to include a cargo carrier on top, and I had our dog and cat. My wife and kids were going to fly to Fairbanks after spending time with family on the east coast. I had a little less than 10 days to get there. I needed to be in Fairbanks in time to pick them up from the airport when they arrived. I completed the trip in eight days.

In hindsight, it was awesome... so awesome I would do it again in a few years when I went back to Alaska. At the time, though, I was seriously questioning my life's choices. It is a really long drive!

In the summer of 2006, vehicle GPS systems weren't really a thing, and I didn't have Google Maps on my flip phone, so I used the good old fashioned paper map. It was a map book — multiple maps for multiple states and two countries (U.S. and Canada). The cat and dog got sick... multiple times. I got held up at the Canadian border and was forced to unpack my car because the Canadian customs agents thought I was transporting personal weapons... I wasn't.

I stayed in shitty hotels since I had the pets with me... it's amazing I didn't get lice. And I got a flat tire just as I crossed the border into Alaska. All that said, it was an adventure of a lifetime. The scenery was tremendous, and I saw a ton of wildlife — rams, bears, moose, etc. My only regret is that I didn't spend more time talking to some of the people I met along the way... but then again, talking to strangers isn't really my thing.

When I arrived in Alaska, I checked into post lodging, washed my truck (I slaughtered at least a million bugs on the trip, all prominently displayed on the grill and windshield of my vehicle), and I met with the battalion rear detachment leadership to figure out what was going on. You see, about six days into the trip, I heard on the news that President Bush was going to extend several units in Iraq as he began to "surge" forces into theater to combat the growing insurgency.

To surge forces, he had to extend certain units already there to cover the gap and buy time and space for the build-up of additional combat power. As it turns out, the 172nd SBCT was one of those units, along with the battalion I was going to take command of, 2-1 IN. After some back and forth, the change of command was on hold until the unit returned from Iraq, which would be in about four months — December 2006.

The period from August to December 2006 was tough for a lot of reasons. Personally, I was disappointed I wasn't taking command until December. I also wanted to deploy forward but didn't. The new chain of command was already in Fairbanks — the incoming Brigade Commander, the other battalion commanders, some of the field grades. We were going to be in command for three years, so the plan was to go into command together and come out together. Aside from the Brigade Commander who changed out a few weeks before the rest of us, that's what ended up happening... but more on that later.

The good news is that we were able to build relationships

with one another, the community, and the division chain of command. The bad news was that we had no command authority. Quite simply, we were not in command, but we were almost like a shadow chain of command. New Soldiers arrived every day. These Soldiers were going to backfill the returning Soldiers who would PCS or ETS upon return.

In short order, the post was bursting at the seams with thousands of Soldiers fresh from basic and advanced individual training (AIT) who were now in Alaska in the early winter. Finding things to do and keeping them busy was a challenge. Discipline and accountability quickly became problematic.

More problematic and distressing, however, was that we still had a unit forward deployed in combat, and they were taking casualties. Families were in limbo. Given the planned August redeployment, some families left early. Others had already shipped household goods and now had nothing. If the unit was coming back in December and then families would PCS, how would this work for schools, spouse jobs, etc.? We also had Soldiers who had redeployed to Alaska and had to go back to Iraq. To the Army's credit, they surged assets to help manage all of this, but it was an intense period of sacrifice and hardship for Soldiers and their families.

Finally, in December 2006, the entire Brigade was redeployed and back in Fairbanks with a change of command scheduled for mid-December, and by the way, the unit was going to reflag. The Brigade would become the 1st Stryker Brigade Combat Team, 25th Infantry Division, and 2-1 IN would reflag to 1st Battalion, 24th Infantry. I was not consulted on this matter, but if I was, I would have advised the decision makers to delay this.

The plan was that during the actual change of command ceremony we would case the old colors, uncase the new colors, and change patches (pull the old patch off the velcro sleeve and put the new patch on). And that's exactly what happened. The problem is that you just tangibly tore away the unit that many Soldiers identified with... a unit they spent 16 months in combat with. While building unit identity would be challenging for me, it was the least of my issues when I took command.

Chaos... Taking the Colors

Within weeks of the unit redeploying, I took command in early December 2006 at the Carlson Center in Fairbanks, Alaska. I was now Legion 6. It was a mass change of command. In other words, all seven battalions changed out at the same time, as well as the Brigade Command Team. In actuality, I took command of 2-1 IN and then shortly after the change of command — during the same ceremony — my Command Sergeant Major (CSM) and I cased the 2-1 IN colors and uncased the 1-24 IN colors.

I hit the jackpot with my Battalion CSM. Taking nothing away from anyone else, Ray F. would turn out to be the best Soldier with whom I ever served. During our time together as a command team, he made me a better leader, officer, commander, and person. To this day, he pushes me to be excellent. The quick backstory on how Ray and I met was during that period from August to December when the Brigade was extended in Iraq, we worked together. Ray was the United States Army Alaska (USARAK), or Division, Operations Sergeant Major (SGM). We met during a planning exercise and connected immediately. Originally, the 2-1 CSM was going to stay when I took command, but he received orders and PCS'd shortly after the unit redeployed. I had a significant vote in the selection of my Battalion CSM. Ray was already waiting on a Battalion, so it was an easy choice and didn't require much bureaucracy or logistics. It turned out to be the best thing that could have happened to me as a battalion commander.

To say that it was chaos when I took command would be an understatement. While Legion 7 and I were a great command team, for a while, it was touch and go. Like, literally, Ray and I thought we would be fired. Given the battalion's extension in Iraq, "Stop-Loss" (which put an end to normal PCS and ETS out of the unit), and the influx of new Soldiers who were still sent to Alaska to replace the Soldiers leaving (who were under "Stop-Loss"), it was mayhem. In short, Soldiers and families kept arriving at Ft. Wainwright, in the middle of an Alaska winter, but no one was leaving — or had left yet.

With an authorized personnel strength of roughly 700, we had over 1,000 Soldiers currently assigned to the Legion. There wasn't enough barracks space, so Soldiers tripled and quadrupled up in rooms designed for only two Soldiers. We converted offices and dayrooms to barracks spaces for Soldiers to stay. It was December in Fairbanks, so the weather was well below freezing every day and it was dark... very dark.

The environmental conditions of a Fairbanks winter — extreme cold hovering in the -10 to -30 range and long, dark days with only three to four hours of daylight — dramatically impacted Soldier, leader, and family morale and energy. Everyone was tired... very tired. And we weren't training. The unit's equipment wasn't back from Iraq, so outside of physical training (PT), there was no structured training going on. And we were in the middle of an identity crisis, having just reflagged from 2-1 IN to 1-24 IN.

At one point, things got so bad that the Division Commander sent a Memorandum for Record to all battalion command teams. The memorandum was hard copy only — it did not go out over email. It essentially said, take back your barracks, reestablish good order and discipline, or you will be fired. I recall very vividly sitting around the small round table in my office with Ray. We each had a copy of the memorandum in our hands. We looked at one another and both of us agreed that this was not how we saw battalion command happening when we dreamed of being in this position. But we needed to act and exercise leadership... like now!

THE LEGION STANDARD: PREPARING A BATTALION FOR WAR

So how did we do it? How did we instill good order and discipline? How did we build a lethal fighting force? Poor morale, overcrowding, substandard living conditions, and not enough to do for all the Soldiers led to misconduct, indiscipline, and, in some cases, criminal behavior. At one point, we had a rifle company leading the Army... the Army... in drug offenses.

It was not a good look; and the Division Commander made that clear to us. As mentioned, Ray and I thought we would be fired for sure if we did not get a handle on this. Over time, the normal rotation of Soldiers out of the Battalion settled some of our overcrowding problems, but there were primarily three things that helped get us on track.

Accountability: Quite simply, we had a one and done policy. We chaptered (separated) Soldiers who committed drug offenses. Some went to jail for a short period of time, depending on the egregiousness of the offense. We were public about this with the message being that we would not tolerate such behavior. However, we did treat each situation on a case-by-case basis. As an example, in one instance, we did not separate a Soldier based on the recommendation of the Soldier's chain of command. This turned out to be the right decision as this Soldier ended up serving a long career in the Army!

We instituted a program called "Alternatives to Drugs and Alcohol." Platoon Leaders and Platoon Sergeants were responsible for developing an off-duty training schedule — primarily in the winter — that afforded Soldiers the opportunity to snowmachine, cross-country ski, downhill ski, ice fish, ice climb, snowshoe, etc.

We wanted to make the access to these types of events easy, get Soldiers out of the barracks, and encourage them to engage in healthy activities instead of binge drinking and/or doing drugs. An added benefit was building greater platoon camaraderie. While not a perfect program, it did make a difference.

Leadership at the Lowest Level: We worked hard to put the right leaders in charge all the way to the lowest level. This extended from Company-level down to the fire team level. Our selection criteria were not complicated — we needed leaders. We could grow technical and tactical competence, and over time we could grow leaders, but in the immediate term we needed people to step in and take charge.

Our subjective assessment was based generally on a Soldier's level of fitness, command/leader presence, personal example, and values. We were looking for the "best athlete." We relied on company commanders and first sergeants to make the right choices. This was not an exact science, and in some cases, we swung and missed. The downside was that this created some turbulence in the formation as we moved leaders around. On the other hand, it was important to get the right leaders in place early, and before we deployed.

The result was that we sometimes had very junior Soldiers leading fire teams, and even squads. As an example, we had a Private First Class (PFC W.) leading a squad of eight other Soldiers, most of whom were either of the same rank or junior to him. While the execution of the tactical task at the time was not perfect, PFC W. provided much needed leadership. When Ray and I ran into him, we decided that we could not have a PFC leading a squad, so we made PFC W. a Corporal

(CPL) and pinned an Army Achievement Medal on his chest for good measure. As an aside, I ran into CPL W. several years after this. He was a Sergeant First Class, Ranger-qualified, and a Platoon Sergeant in the 101st Airborne (Air Assault) Division.

Operation Darby: We went to the field. It was clear to Ray and me that we needed to get the Battalion out of the barracks and the garrison of Fort Wainwright and get them to the field. We took the Battalion to the Donnelly Training Area (DTA) in March, several hours from Fairbanks. This afforded us the opportunity to exercise several things, not the least of which was to develop mental and emotional toughness in an Alaska winter and to hone our Arctic skills. There were several challenges we faced to make this happen, but these challenges became opportunities... opportunities to problem solve, lead through adversity, and train to standard.

We intentionally chose Operation Darby as the name for the field training exercise because we wanted to focus on squad level tactics (Darby is the first phase of Ranger School and is centered on squad tactics). In many ways, Operation Darby became the turning point for the Battalion.

Equally important to all of this was how to deal with the reflagging of the Battalion. To recount this in a little more detail, for several reasons above my paygrade, the Army officially redesignated the Brigade from the 172nd Stryker Infantry Brigade Combat Team to the 1st Stryker Brigade Combat Team, 25th Infantry Division. All the Battalions were redesignated as well, with 2nd Battalion 1st Infantry Regiment reflagging to 1st Battalion 24th Infantry Regiment.

What this meant practically was that the patch we wore on our left sleeve changed and the Battalion colors and unit guidons also changed to reflect our new unit — the old colors/guidons were "cased" (or covered) and the new ones were "uncased" (or uncovered). Below the surface, it was the intangibles that didn't change. Soldiers who fought, bled, and buried their buddies under the colors and insignia of one unit were told to leave that behind and now re-align under an entirely different unit. The transition was not easy.

The sound of the tear of Velcro-on-Velcro was deafening across the convention center as over 4,000 Soldiers removed the 172nd patch and replaced it with the 25th Infantry Division patch. And just like that, the Soldiers of 2-1 IN were now in 1-24 IN.

Unfortunately, the by-the-book, clinical, and unemotional manner by which we approached the unit's reflagging was demoralizing for many of the Soldiers and leaders. Unintentionally, we announced to them that the former unit is in the past and that it no longer matters. Many were angry, sad, and frustrated. The good news is that the unit is the people, not necessarily the patch, so we could overcome this.

As Ray and I thought about how to approach the situation, we had some flexibility in selecting our Battalion motto. This turned out to be a blessing in disguise as we were able to use this as an opportunity to shape the culture of the formation right from the start. We wanted to retain the greatness of both units and to honor the Soldiers who went before us. "Deuce Four" (1-24) had their own great history. 2-1 IN was known as the Legion.

In short, unit history matters. We became "Deuce Four — The Arctic Legion". Our motto was "Strength and Honor." We took this from the movie, *Gladiator*. It was the exchange / greeting between the Legionnaires of the Roman Legion. For us, it became the anchor around how we wanted to lead, train, and fight as a Battalion. Strength and Honor were the core values of the Arctic Legion. Regardless of the patch on our left shoulder, the Soldiers were the Legion. This was our culture.

Soldiers are the unit. But there is also history to the unit, and in this case, specifically 1st Battalion 24th Infantry. The Regimental Crest was designed for a certain reason. "Deuce Four" existed for a reason. And Ray and I were well aware that the 1st Battalion was the only active Battalion in the 24th Infantry Regiment. With all of this in mind, we made sure to include Colonel (Retired) John Komp in all these discussions. COL Komp was the Honorary Colonel of the Regiment. He served in World War II, Korea, and Vietnam. During the Korean War, he was a company commander in the 24th Infantry Regiment.

At one point, I was talking with COL Komp on the phone, and I was sharing my mailing address so he could send along some books on the Regiment. I lived on Bastogne Court at the time. As I began to spell out "Bastogne" for him, COL Komp stopped me and told me that he knew how to spell it. He was there! What an honor to have COL Komp as our Honorary Colonel. He visited the unit often and supported us in all that we wanted to do. COL Komp passed away on 31 December 2015 at the age of 93.

ON THE ROAD TO WAR

Operation Darby (which happened about 3.5 months into command) was the turning point for the Legion. We knew early on in command that we were going to deploy back to Iraq, so this was an important point for us. As Ray and I pushed the Battalion forward, we wanted to (1) build combat teams and (2) develop combat leaders. Fortunately, we had very strong company commanders and first sergeants, and field grade officers who understood training management and logistical resourcing to support our intent. Operating with my intent and vision, these leaders were able to move the Legion forward.

Excellence in combat is achieved before you cross the line of departure. There were four tenets to this that guided our philosophy in building lethal combat teams in the Legion: (1) Physical Fitness and Mental Toughness; (2) Marksmanship; (3) Maneuver Live Fire Training; (4) Care of Equipment. We wanted the Legion to be the toughest unit in the Brigade — physically, mentally, and emotionally. We agreed early on that we wanted to be Ranger Regiment-like in our adherence and enforcement of standards and our high training expectations.

While we may not have had the budget to support training at the level of the Ranger Regiment, the goal was to be like the Regiment — the gold standard for Infantry formations in the Army.

Physically, everything we did was with an eye toward being combat ready. Battalion level physical training events were conducted in full kit to replicate the conditions of combat, where we would need to wear helmet and body armor for 12-18 hours per day. "Lead by example, from the front" was Ray's guiding leadership principle and he did just that. Soldiers and leaders were inspired by his example of fitness, toughness, and exceptional Soldier skills. I was inspired!

We also planned and executed Mangudai events for the leader teams (PL/PSG, CO/1SG, BC/CSM) to foster team building and cohesiveness. The term is believed to be inspired by the rigorous selection process for elite warriors under Genghis Khan during the Mongol Empire. The Mungadai concept was conceived by Maj. Gen. David Grange, an Infantry officer with many years of service spent in elite special operations formations. The Mungadai consists of events that test the teamwork, physical stamina and leadership abilities of senior officers and NCOs under tough conditions to build esprit de corps and team cohesion.

Units doing hard things together are more cohesive teams. Our training focus was to drill down on the basics, beginning at the fire team level. Teams that take care of their equipment and maintain it properly, shoot well, and can execute under live-fire and near combat conditions make for better squads, better platoons, better companies, and a better battalion.

We believed that what separated good combat leaders from average ones were (1) aggressiveness, (2) being on point, and (3) knowing the enemy. The first two tenets were about mindset. I learned early on in my career that in combat it was important to maintain an offensive bias to gain and maintain the initiative, while doing this within our values. We could not be anchored to our combat outposts, which tied to the second point.

Leaders needed to be out in the area of operations to understand the people, the enemy, the terrain — leaders had to "feel" the battlefield. To the third tenet, we built a robust Leader Development Program focused on understanding the environment, the history, the political and cultural dynamics, and the enemy of the area of operations where we would fight. We also conducted leader team training that focused on various combat vignettes where our platoon and company level leaders discussed how and why they would make certain decisions, often dictated by enemy action or inaction.

If there was one leader who represented the mindset of the Legion, it was Staff Sergeant (SSG) Eric H. Eric was the prototypical 11-Bravo Infantry Non-Commissioned Officer — hard, demanding, and deeply protective of his Soldiers — and he was the epitome of the combat leader we wanted in the formation. Eric was a Squad Leader in Charlie Company. Already a proven combat leader, he absolutely internalized what Legion 7 and I were trying to build in the battalion. An exacting, standards-based leader, he led from the front. If his Soldiers wanted to know what right looked like, all they had to do was watch him.

My fondest memory of Eric was when I ran into him and his squad at the top of Birch Hill — the on-post ski slope at Ft. Wainwright — one cold, frigid Fairbanks morning. It must have been closer to the spring since I recall plenty of sunlight reflecting off the snow on the ski hill. I was up there doing physical training with another squad. We had just snowshoed to the top, and we were catching our breath before moving back down.

SSG H's squad literally appeared out of the woodline at the top of the hill. They were on the infamous "white rockets," the Army issued skis that could double as cross-country skis or downhill skis. They were multi-purpose and were not necessarily conducive to first time skiers, especially those wearing helmets, body armor, and with slung rifles (or in this case, rubber ducks — i.e., inert rifles).

Always training, Eric believed in learning by doing and his young Soldiers were ready to learn. While Eric had a few years in Alaska under his belt and had spent some time on "white rockets," he was by no means a master. But true to form, he led by example. The hard-charging, take-no-prisoners Squad Leader led his Soldiers down the hill. At some point, it became a yard sale with skis and poles littering the mountain, but they eventually made it, stowed the skis on their backs at the bottom, and then foot-marched roughly six miles back to the company area.

An incredible leader, he made sure that his squad was laser focused, trained, and that they took care of one other. Unfortunately, despite our best intentions, Eric and I lost contact with one another after the deployment and the battalion change of command. The good news is that I reconnected with Eric a few years ago. I'm a better man for it!

HINDSIGHT IS 20/20

As I reflect on my time as the Legion commander, there are two shortfalls that immediately jump out. First, I did not prepare the unit (or myself) for the operational tempo of combat. Having deployed before, I knew what it would be like, but I did not even consider this in our ramp-up to war. In fact, in some ways, I probably pushed too hard in our pre-deployment workup.

I do not believe that the Battalion was stressed or tired by the time we deployed. On the contrary, we were well trained and razor-edge sharp in our warfighting skills. That said, we hit the ground running and never stopped until we redeployed 12-months later.

Ironically, I would often tell my leaders that they "needed to take an appetite suppressant" when planning training. The desire to leave no stone unturned in our training workup would often mean that we wanted to fill all the white space on the training calendar. The bill payer was time off, family time, time to take care of personal business, or time to just sit around and relax.

Fighter management — how we managed rest and refit while deployed (especially with our leaders) — was not something we discussed at any point that I can recall. I know personally that I was exhausted by the time we redeployed at the end of 12 months. The lesson here is that to stay switched on 24/7 for 365 days is quite simply not possible, and that over time the inability to manage rest and refit degrades the fighting capability and performance of a unit. Leaders need to account for this.

While you are in theater, it is more difficult to build in rest and refit periods. The pace of operations and mission requirements will dictate the terms. Prior to deploying, however, it is not weak or soft to give Soldiers and leaders time off. A few years later as a Brigade Commander, I learned my lesson. When we were not deployed or in a training cycle, I would try to make every three-day weekend a four-day, as an example, or ruthlessly enforce the early release for family time every Thursday or Friday.

The second shortfall was my inability to be a good team player. While I do not believe this was catastrophic in any way, I was more concerned about what was best for the Legion and not the larger organization, i.e., the Brigade... and by extension, the Division. I had a hard time seeing (or refusing to see) the larger picture at times, and how the Battalion fit into this.

If the task, mission, or operation was not best for the Legion, I pushed back on my commander. I was more of a pain-in-the-ass for the Brigade Commander and the Brigade Staff than I needed to be. Often this was over very minor things like range usage or details and taskings. Looking back, I am embarrassed by my behavior. The irony is that a few years later, I became the Brigade Commander, and I would find myself asking the Battalion Commanders how a particular training event or action made us a better Brigade — to think beyond just their Battalions.

From September 2008 through September 2009, the Arctic Legion excelled in a complicated and complex area of operations (AO) in Western Diyala Province, Iraq. Our AO was defined by a thinking and adaptive enemy. A defining operation for us was a Battalion (-) air assault to South Balad Ruz to conduct a clearing operation against pockets of enemy resistance. From the time we received the mission to execution, the Legion had to turn on a dime. It was a three-day operation that required us to maintain an active presence at each of our combat outposts with enough combat power to project strength, while executing an air assault with the main effort of the battalion to find, fix, and finish enemy resistance in a historic enemy stronghold. This required split command and control, competing resources, and hasty planning, flexibility, and execution on the part of Soldiers and leaders. The only reason we could do this was because we were prepared. We put in the work before we deployed.

I played in the Super Bowl. The Soldiers of the Legion represented everything that is great about the American Soldier.

Commanding an Infantry Battalion in combat is the dream of every Infantry Officer... and I had the honor of doing that. I loved the Legion and our Soldiers, and I love them today. Not a day goes by where I do not think about those days and the lessons I learned about leading Soldiers. To command, lead, and serve the Soldiers of the Arctic Legion was a gift. I am better because of them. STRENGTH AND HONOR!

6. BRIGADE COMMAND

IN THE BEGINNING

Coming out of Battalion Command, I felt that I was competitive for Brigade Command. I had three years of Battalion Command, took the battalion to combat, strong Officer Evaluation Reports (OERs), three key and developmental jobs as a Major with equally strong OERs in those jobs... I felt pretty good about my chances. My branch manager seemed to think so as well. I was already selected for the Senior Service College and had deferred twice. There were lots of moving pieces to my next assignment and my future, the biggest of which was that our middle daughter was a junior in high school.

I gave up command in December 2009. December is an odd time to change command and out of the normal PCS/personnel rotation season. I submitted a high school stabilization to stay in Alaska one more year so my daughter could graduate high school. The good news was that I was being tracked as a high performing former battalion commander. The bad news was that there were no jobs for me in Alaska and I also had to do a utilization tour for my graduate school and Ph.D. That meant we were moving. The Infantry Branch Chief denied the high school stabilization. I was officially a shitty father. I moved my daughter for her senior year of high school.

Long story short, I went back to the United States Military Academy for a utilization tour. There was no way the Army was going to send me for a Ph.D. and not get some use out of me. Given that I was initially hired as a rotating Ph.D., I should not have been surprised. I left my family in Alaska to finish out the school year and I moved to West Point in January 2010 to catch the second semester at the Academy. In June, I went back to Alaska to move my family from Fairbanks to West Point.

My initial plan was to stay at USMA and serve out my time in the Army and retire. Even though I felt that I could be selected for O6/Brigade-level command, I was thinking hard about stabilizing and the next life steps.

In my mind, this plan made complete sense. I was already over 20 years of service. The problem was that there were no available permanent positions in the Department of Behavioral Sciences and Leadership that would allow me to stay and retire at 30 years, the mandatory retirement date (MRD); and as an Infantry officer I likely would be PCS'd in two years, and I still had to go to Senior Service College. This is a long way of saying that we decided to compete for Brigade Command.

Shortly after pinning on the rank of Colonel in October 2010, I found out that I was selected for Brigade Command. Fort Wainwright and the 1st Stryker Brigade Combat Team, 25th Infantry Division (1/25 SBCT — Arctic Wolves) was our first choice, and we got it. We were going back to Alaska and the unit I just left.

First, however, I had to go to the Senior Service College. The Army, in its infinite wisdom, selected me for the National War College in Washington, D.C., which meant I would move once again. This would mean a move from Alaska to West Point in June 2010; a move from West Point to D.C. in June 2011; and then a move back to Alaska in June 2012 — three moves in three years. Fortunately, mentors and senior officers weighed in and I was re-slated for a Senior Service College Fellowship at Teacher's College, Columbia University which allowed me to stay at West Point and commute into New York City. And it turned out to be an incredible fellowship.

The Arctic Wolves were coming off a deployment to Afghanistan in the spring of 2012. I was slated to take command later in the summer. I received a call early in March 2012 that I would be accelerated into command and take the colors in two months — in May 2012. I still had several requirements to complete for the fellowship, and a move in early May wasn't optimal for moving my family, given that my son was still in school. We eventually got it all sorted out and on 16 May 2012 I took command of the Arctic Wolves from COL Todd W.

THE WOLF

The enormity of what I was stepping into as Wolf 6 didn't really hit me until we were "inspecting the troops" as part of the change of command ceremony. This element of the ceremony required me — with the outgoing commander and the commanding general who was officiating the ceremony — to literally walk around the entire formation. This included all seven battalions — 1-24 IN, 1-5 IN, 3-21 IN, 5-1 CAV, 2-8 FA, 25th BSB, 25th BTB — and associated companies and platoons. In total, this was over 4,400 Soldiers. Holy Shit! Was the Army entrusting me with this responsibility — as it turns out, it was!

At the time, my Command Sergeant Major (CSM) wasn't on the ground. Ray L. was still at Ft. Riley in his current assignment. He would be joining the Command Team in a few weeks. So, Craig S. was my acting CSM — or Wolf 7. Craig was the longest serving CSM in the Brigade and he was currently Automatic 7, the CSM for the artillery battalion, 2nd Battalion 8th Field Artillery (Automatic). Quite simply, I could not have asked for a better senior NCO to help me onboard and teach me the ropes. While I obviously had three plus years of time in Fairbanks and at Ft. Wainwright as a Battalion Commander, being the Brigade Commander was a completely different beast!

Prior to assuming either LTC-O5 or COL-O6 command, the Army sends you to the Pre-Command Course (PCC). All in all, I found the course prior to both Battalion and then Brigade Command to be very helpful and informative in preparing me for the opportunity ahead. You also have access to the Army's most senior leaders who often came in person to talk to us at Ft. Leavenworth, Kansas, where the course was held. This is an indicator of the priority that the Army as an institution places on the course, as well as the enormity of the responsibility that we all have as Battalion and Brigade Commanders.

At PCC for Brigade Command, then-LTG David Perkins spoke to us about his view on leadership and the roles and responsibilities inherent with O6-level leadership. This included a thought or two on leading an organization with subordinate battalion commanders who were all Department of the Army Centrally Selected to be battalion commanders and were only an assignment or two away from where we were as Brigade Commanders.

In other words, they were a select group of highly qualified officers. His suggestion was that upon taking command, within the first 24-48 hours, meet collectively with your Battalion Commanders — just you and them. So, I did that and followed his advice to a tee.

It went something like this. We all assembled in my office — Matt M., Jason W., Scott S., Erik K., Tom R., Mike S., Mick B. and me! We sat around my conference room table. There were not a whole lot of pre-existing relationships. This was really the first time most of us would be working together. Jason and I worked together at Ft. Hood several years prior and deployed to GTMO together as part of JTF-160. A couple of the other guys knew each other tangentially, but it was really like a bunch of dogs meeting for the first time at the dog park... some strutting around, ass sniffing, trying to figure out who is who… and all wondering what I was all about. What was I going to be like as their boss?

With this context, I led with an opening question: "Who is the best battalion commander here?" As you might imagine, it was crickets. Guys were looking at their feet, staring blankly at the table or their open notebook; anything to avoid making eye contact with me or each other. After some awkward silence, I responded with: "Clearly, I am. Otherwise, I wouldn't be the Brigade Commander."

While this didn't quite have the gut-busting laughter that LTG Perkins got when he shared the story at PCC, my version did receive some chuckles and a few "where the fuck is this going" looks from the commanders. I continued with something along the lines of: "I was a battalion commander and at least in the Army's eyes a pretty decent one since I ended up being selected for Brigade Command. My point, though, is that I am not interested in being a battalion commander again. You command your battalions. I'll focus on the Brigade."

And with that, I set what I felt was a tone for our time together. As battalion commanders, they would have the autonomy to command and lead their formations. We shot the shit getting to know one another, I elaborated on this a little more, and that brought to an end our first moments in command together.

For over two years we commanded the Brigade together. They were a great group of leaders and commanders. We all had our strengths and weaknesses, but there is no doubt that we left the Brigade better than how we received it. All talented in their own rights, I greatly appreciated their counsel, wisdom, honesty, and humor. Mick B. was unique in that he was a Special Forces officer and had served in the Brigade previously during the deployment to Afghanistan. While we were not peers in rank, he became a trusted agent and a good friend. He was a confidant who I could bounce ideas off of, and I knew I would get honest and unvarnished feedback. I believe every commander needs someone like Mick.

As I was working on this book, it was announced that General Charles Jacoby, U.S. Army (Retired), passed away on 1 April 2025. Over the past couple of days, I've been reflecting on this. I knew him pretty well. He was the United States Army Alaska (USARAK) Commander when I was a Battalion Commander, and then later he was the United States Northern Command (NORTHCOM) Commander when I was the 1/25 SBCT Commander.

Given the Arctic Wolves' Homeland Defense responsibilities, GEN Jacoby visited Alaska at least twice to see the Brigade and receive an update on our mission. I last talked to him when I was the Brigade Tactical Officer (BTO) at West Point, and he was the Modern Warfare Institute's (MWI) Distinguished Chair. Over the span of a career, my interactions with him were somewhat brief, but very impactful.

Within about a month of taking Battalion Command, I was pretty sure he was going to fire me and the rest of my peers. I wrote about this earlier. To say that misconduct in the battalion (and across the Brigade) was out-of-hand is an understatement. In my defense, I had roughly 1,100 Soldiers assigned against my authorized approximately 700 Soldier-strength. It was pure chaos, and it took everything within the powers of the battalion leaders down to the team level to maintain control of the madness.

But all of that is an excuse...we were failing! Somewhere I have the memorandum then-MG Jacoby sent to the battalion command teams directing us to "take back your barracks." The memorandum was for our eyes only and was distributed in hard copy. No email! But it had the right effect.

So that's what we did, we took back our barracks. And as the Commanding General, MG Jacoby supported us. He underwrote the risk as we took the battalion to the field during reset and spent four weeks at the Donnelly Training Area (DTA) in March 2007 for Operation Darby, focusing on the basics at the team and squad level. He put money and resources into our fitness facilities so we could maximize physical training time when the Alaska weather forced us inside. And he underwrote the risk on the Legion School of the Soldier, a program designed to instill accountability, standards, and discipline.

When he was the NORTHCOM Commander, I recall having dinner with him in a Fairbanks restaurant after a day of briefings and troop visits. His purpose in coming to Ft. Wainwright was to get an update on our Homeland Defense Mission and Posture. It was a bit of a tough time for the Brigade as we just were off ramped for Afghanistan.

Instead, we pivoted to the Pacific Theater through exercises aligned with the Army's Pacific Pathways Training and Exercise program; deployments to Korea in support of the "Go to War" mission on the peninsula; and then our defense of the Nation mission in Alaska.

By tough time, I mean there was the excitement and build-up for Afghanistan, and then to come down off that was rough on morale. It was a bit of a letdown. But GEN Jacoby didn't let us feel sorry for ourselves. He also didn't blow smoke up our asses. He made sure we understood the importance of our mission set and that as professionals we needed to focus on the priorities the Army and the Nation had set for us.

When I last spoke with him, we talked about his unconventional career — he taught at the United States Military Academy as a junior officer; he went to the School of Advanced Military Studies (SAMs); he commanded an unconventional brigade (JTF-Bravo), an unconventional division (USARAK), and an unconventional corps (I Corps). We talked about the fact that we need good people at West Point, training and educating the future of our Army. We talked about "growing where you are planted."

You don't need to serve in the Ranger Regiment or Special Operations to be a great Infantryman. And he was a great Infantryman. You don't need to be professionally connected to get promoted. Do the jobs you're given and do them well. As he said to me, and then later reiterated in his retirement speech, he loved being a Soldier. And he loved Soldiers. Until Valhalla! Rest in Peace, Sir.

THE BIG 5

For many reasons, Brigade Command is a lot different than battalion command. Where battalion command is that perfect intersection of exposure (to Soldiers) and experience (professional), with brigade command you have more experience, but less exposure. Given the seven battalions and 30+ companies in the Arctic Wolves, I had to trust more than ever that the leaders at every level understood my intent and were willing to execute that intent. Brigade Command is not the time for micro-managing. Much of my focus was up, out, and across. As much as I wanted to spend more time "down," my relationships with the Division, Corps, Garrison, and adjacent units were critical to resourcing the Brigade for success. Given this, I had to trust my battalion command teams and make sure the other members of the "Big 5" were empowered to make decisions on my behalf and for the Brigade.

The "Big 5" typically refers to the Brigade Commander, the Brigade CSM, the Brigade Deputy Commander (DCO), the Brigade XO, and the Brigade S3. Given that many of my responsibilities took me up and out, I had to make sure that my other teammates in the "Big 5" were empowered and understood my intent. Shared and common understanding was critical for us to effectively run the organization. My personal style was such that I communicated and was accessible — maybe too accessible at times. We talked often as the Brigade Leader Team and made sure we spoke with one voice. It also helped that we were all very good friends.

Although I was not new to Alaska, I was new to Brigade Command and fortunately I had a Command Sergeant Major who already had some time at the Brigade level. Ray was coming from the 1st Infantry Division where he had already served as a Brigade CSM. He knew the ropes and I listened. It hurt to lose him early to be the Division CSM of 10th Mountain Division, but it was great for Ray and the Army.

Bill, Jason, and Jimmy also brought a wealth of experience and knowledge, having served with the Division and the Brigade previously. They knew people and systems. Super talented, their counsel was invaluable. All five of us stayed together for the two plus years I was Wolf 6. I struck gold with these guys!

There were a lot of moments that I think really defined us as a command team, but one for sure... and humorous... was the annual calendar synchronization meeting to figure out who was going hunting and when. With moose season in September and a Caribou trip up north shortly after, we had to keep our priorities straight. What was the hunting plan... and oh yeah, how would we cover down on the leadership of the brigade while we were hunting.

What we decided was that Ray, Bill, Jimmy, and Jason would go on a caribou hunt while I stayed back and kept the brigade running. They would assume the helm while I went moose hunting. I can't do justice to the caribou hunt they went on and will let them tell that story! However, I can share the story of my moose hunt.

I planned my moose hunt over a four-day weekend. I left

on a Thursday and came back the following Monday evening. My best friends in the whole world are Mike and Peggy, and by extension their wonderful sons and families. I went out on this hunt with Mike and one of his sons, Jerry. I don't want to describe the location in too much detail so as to not give away the spot, but Mike and I arrived at moose camp on a Thursday afternoon. We immediately took to scouting out locations to spot and shoot a moose, and this involved climbing trees... very, very high trees.

I am not afraid of heights, and I climbed trees as a kid... city or suburban trees... not skyscrapers like the trees in central Alaska! Mike, on the other hand, grew up in Sitka, Alaska and he and the trees were one. Not to be shamed or outdone by Mike (who is more than a few years older than me by the way), I did not let on about my hesitation... and so I climbed. All was good, but thankfully we were coming to the end of the day. We had scouted enough spots to get a good sense of where we wanted to go the next morning. We decided to check out one more spot before we headed back to moose camp for some brown water and dinner over an open flame.

So up we went. Mike went first. I slung my rifle across my back and started climbing. About maybe 50 feet up, Mike yelled down to me that this was a good spotting tree and asked me what I thought. As a total amateur in assessing spotting locations, I of course concurred! And then we began to descend. At some point on my way down, I stepped on a branch that snapped, and I began to fall. The tree branches were breaking my descent... fortunately... and I managed to grab a hold of a branch that stopped my fall and swung me into the trunk of the tree. As I like to say, "my monkey strength saved my life!" While this may be a bit of an exaggeration, the fact is that I stopped falling and climbed down the rest of the way.

When Mike came down, I was sitting on a tree stump. Mike thought I had fallen all the way to the ground and expected to see me laying at the bottom of the tree in some state of semi-consciousness! So, the fact that I was sitting on a stump and trying to catch my breath was a good thing.

After a few minutes, we started to head back to moose camp. Sore and a little short of breath, I otherwise felt OK. The next morning was a different story. I could hardly breathe, and it felt like someone was pushing on my chest. Against conventional wisdom and Mike's sage advice, I decided to stay through Monday evening instead of going back and seeing a doctor. Surviving on brown water and aspirin, we hunted and fished over the next several days.

When I finally got home that Monday night, I called the Brigade Surgeon and asked if he could meet me at my office in the Brigade Headquarters. Rich was a great dude, and he agreed to meet me there in about 30 minutes. At my office, I told him what happened, and he checked me out. I broke two ribs. This was not good as there was no magic pill that was going to fix this. Depending on who you ask, this story went in a couple directions from here.

Given that I'm the one telling it, I'm going with my version. I toughed through four days of hunting and fishing in the Alaska wilderness with two broken ribs and then proceeded to continue to PT and lead the brigade while the ribs healed, with no one knowing any differently. Others will tell you that the legend of "McRib" was born! Either way, it's just another day commanding an Infantry Brigade Combat Team in the Last Frontier.

THE BRIGADE MISSION

I never had the opportunity to deploy the Brigade to combat. We were on again/off again for Afghanistan. At one point we were set to deploy, but that would have involved only a portion of the Brigade given the mission and force cap constraints, with most of the formation remaining in Alaska. In many ways, I'm glad that didn't happen. Instead, we were able to pivot our attention to the Pacific, focus on the NORTH-COM mission set, train and refine our Arctic skills, and conduct a rotation at the National Training Center.

I spent quite a bit of time in Korea supporting various exercises and priorities for the United States Army Pacific (US-ARPAC) Commander as it applied to the Korean Peninsula. We did some great Warfighter Exercises that really sharpened our staff processes and command and control capabilities. It was in Korea where I was reminded of a good lesson — keeping your boss informed — and its corollary, if your boss is on the email, he should be in the "To" line, not the "Cc" line.

MG Tom Vandal (d. Oct 2018) was the 2nd Infantry Division Commanding General. He was a great guy, and a great commander. Given the amount of time I spent in Korea working for him, he treated me as one of his own commanders. This included mentorship, tomahawk throws, and an occasional kick in the ass. During one particular exercise, my brigade was executing a counter-reconnaissance mission. This was not a standard mission set for a Stryker Brigade Combat Team. Nonetheless, there we were reporting on enemy movements and providing early warning for the Division.

I always prided myself on how I shared information, being transparent, and communicating early and often. During this exercise, I was pushing reports to the Assistant Division Commander for Operations (ADC-O) and the Division Main Headquarters. For whatever reason — and not a good one — it never occurred to me to include the Commanding General (CG) on these reports. MG Vandal got a hold of the reports at some point, complimented me on the reporting, and then asked what was more of a rhetorical question: "Wouldn't it be a good idea for the CG to have seen these reports?"

To say that he was direct in his delivery would be an understatement. Point well taken. He was right. I needed to inform his decision making and he needed my reports to do that. So, on subsequent reports, I "CCd" him, to which MG Vandal then pointed out — this time more directly — that he was the senior officer in the Division, the report was coming to him, and he should be on the "To" line. I appreciate the lesson, Sir. You were right!

Back in Alaska, the Brigade made the most of our opportunity to fight and win in the Arctic. Given the constant deployments and redeployments to Iraq and Afghanistan over the past 10 years, the Arctic Wolves were losing the Arctic warfighting edge. The battalions flexed their muscles in the local training areas — Tanana Flats, Yukon Training Area, Donnelly Training Area, and Black Rapids — conducting training and exercises that would grow into what is called today the Joint Pacific Multi-Readiness Center-Alaska.

Arctic skills are perishable and by the time I left command, I felt like we had moved the needle in this space. While the Stryker vehicles are no longer in Alaska, I would argue that with the right maintenance, care of equipment, and training, it was a capable Arctic fighting vehicle. This, combined with the light Arctic Warrior — the Soldier — made for a lethal combination.

Most importantly, however, I had an opportunity to groom and develop the next generation of Army leaders. My leader development program focused on the Captains and the Majors, and CSM Ray L. and I ran a joint program for the leader teams down to company level. A highlight for me was a Mangudai event at the Northern Warfare Training Center at Black Rapids where I took all the Company, Troop, and Battery commanders on a several day team building event focused on land navigation, high altitude marksmanship, and mountaineering skills.

I loved the Arctic Wolves. It was a tremendous team. Tragically, though, I lost eight Soldiers while in Brigade Command — and not one was in combat. Soldiering is a dangerous occupation, both on and off duty. We honored each and every one of those Soldiers the way we would have if they died in combat. No Soldier, however, better epitomized the spirit of the Wolf than that of Stephen Stoops.

On 8 January 2012, the enemy was inside the wire. With some downtime, the platoon put down their weapons, put on their workout clothes, and threw together a game of football on Forward Operating Base (FOB) Eagle, an Afghan Army base adjacent to the Americans' FOB Apache. They had played for 45 minutes when an Afghan soldier in full combat gear "came out from behind the bleachers and just started shooting at us," recalled Sergeant Stephen Stoops. Stoops turned to find Private First Class (PFC) John Bolan and Private First Class Dustin Napier down on the field, an oval inside a track like the ones at any American high school.

Stoops sprinted 150 yards across the open field to the base's gate as rounds snapped by his head and ricocheted off the ground near his feet, throwing up small clouds of dust. He grabbed a weapon, and with another Soldier, moved on the assailant. He emptied his magazine. Out of ammo, Stoops ran over and hit the rogue soldier in the face with the machine gun's butt stock until he stopped moving.

Tragically, PFC Dustin Napier died that day. "I want people to remember that [Dustin Napier] was an outstanding soldier, husband, brother, son, and friend that paid the ultimate sacrifice. He was kindhearted and always had a smile on his face," Stoops said. "I will never forget Napier; he has touched my life."

On 23 July 2012, as his Brigade Commander, I had the enormous honor to pin a Bronze Star Medal for Valor on the chest of SGT Stephen Stoops. What drives a person like SGT Stoops to put his own life in danger? Where does this courage come from?

In his own words: "The relationship we have with our battle buddies is that everyone is family... it doesn't matter what someone says or has done. Your [buddy] will always have your back."

Stephen Stoops died on 11 May 2021. Great combat teams are forged through tough training, shared experiences, relationships, and trust. They are family. They have each other's backs. Great combat teams have Soldiers like SGT Stephen Stoops.

My favorite training area in Alaska is the Donnelly Training Area, also known as DTA. There are a lot of reasons I loved DTA. I found range control easy to work with; I liked being away from the flagpole, even when I was the Brigade Commander; and I could exercise almost all the capabilities and enablers of my formation. From a training perspective, you were only limited by your own imagination. But one powerful aspect of DTA was that the beauty and wonder of all of Alaska was on display down there — the cold, the wind, the austereness of the environment... I loved it all.

On one late winter/early spring morning, I was standing outside of a building on a field landing strip where we had established the Tactical Operations Center. It was cold, the wind was still, but the sun was coming up and really illuminating the Alaska range hundreds of miles away. I was drinking a cup of coffee and talking with one of my battalion commanders. We stopped talking and took in the view... it was breathtaking.

I gave up command of the Arctic Wolves on 19 June 2014. While I would always be a leader, this was the last time I ever commanded an Army formation with responsibility for the training, readiness, and health and welfare of Soldiers. I miss it tremendously.

7. FINAL ASSIGNMENTS

CENTCOM...AND THEN THE BRIGADE TACTICAL OFFICER

When I left Brigade Command and was selected to serve in a "black book" assignment, I felt confident that I had a good shot to continue to serve as a General Officer (GO) in the United States Army. Even though I retired as Brigadier General, it didn't exactly work out the way I thought it would. That's a story for another time.

After Brigade Command, I went on to serve another nine years in the Army. From 2014-2016 I was honored to serve as the Director of the Commander's Action Group (CAG) for GEN Austin and then briefly GEN Votel at United States Central Command (CENTCOM). I was definitely at the tip of the spear for some of the biggest and most important strategic and national defense events during that period.

There's not a whole lot I can say about this time, given the nature of our work. I do recall, though, that when I first arrived in Tampa and was learning the ropes of my new job, I felt totally overwhelmed and wondered if this was something I could do. I don't lack confidence, but this was a whole different beast! And while it was relentless, I figured it out, just like any other job I've had in the Army.

I was part of an incredible team of professionals, and they showed me the ropes. I tried to listen more, stay dialed in to what GEN Austin was saying and asking, and sought guidance and feedback from my teammates, both junior and senior to me.

We were Soldiers, Sailors, Marines, and SEALs. As the CAG—special advisors to the CENTCOM Commander—we were well aware of our enormous responsibility and the trust that the Boss had in us. I was essentially joined at the hip with GEN Austin. And in Curtis, Chris, Rich, and BK, I had nothing short of exceptional teammates and friends.

I miss those times. We did travel quite a bit. I joke—although it was true—that my income was tax free for two years because I was overseas the whole time. We spent almost all our time in the region with the Boss—Iraq, Afghanistan, Pakistan, Egypt, Jordan, the rest of the Middle East, Central Asian States, and other points across the globe. We were at war. Not only did I have a front-row seat to leadership and decision making at the highest level, I was a participant.

After CENTCOM, I went back to the United States Military Academy. My plan was to do three more years and then retire at 30-years. Instead, I stayed for seven years. In 2016, when I left CENTCOM, I could have gone to another "black book" billet to serve as an Executive Officer to a 4-Star General. This would have increased my chances for GO selection and would have probably put me over the top. But I felt it wasn't something I wanted to chase, and so when I had an opportunity to select my final assignment, I chose West Point.

Getting assigned to West Point wasn't difficult, especially given that the Superintendent, Lieutenant General (LTG) Robert Caslen, was my former Division Commander when I was deployed to Iraq in 2008. I was excited to work for him again.

The plan was to stay from 2016 to 2019 and then retire at my Mandatory Retirement Date (MRD) in May 2019 at 30 years. I thought I was going to be the Director of the Department of Military Instruction (DMI), essentially in charge of all cadets' military training. Instead, when the Commandant of Cadets, Brigadier General (BG) Diana Holland, called me to welcome me to West Point, she told me I was going to be the Brigade Tactical Officer (BTO).

When I was a cadet, we didn't have a BTO until around 1987, but the irony of the position is that I was not always aligned with the rules and regulations as a cadet, and now I would be enforcing them! Truthfully, I was a little disappointed at first, but in the end, serving as the BTO was an incredible and impactful experience.

I would argue that being the BTO was as close to Brigade Command as you could get without being in command. You lead, you develop, and you train. Organizationally, you look like a Brigade. There are over 4,000 cadets inside the Brigade Tactical Department (BTD). There are four Regiments (akin to Battalions in a Brigade) and then 36 companies — nine per Regiment. Each of the Regiments is led by a Regimental Tactical Officer — a LTC or junior COL — a former Battalion Commander. And each of the Companies is commanded by a Company Tactical Officer — a Captain or junior Major — a former Company, Battery, or Troop Commander.

Each Company has a senior NCO — either a Sergeant First Class or Master Sergeant. There are roughly 120 cadets per company. Over time, I suspect the organization has changed a bit, but this is what it looked like in 2016.

An interesting dynamic in the Brigade Tactical Department is command authority. Important command matters like signing awards or administering the Uniform Code of Military Justice didn't reside with the BTO or the RTOs. Instead, it ran from the Commandant of Cadets directly to the Company Tactical Officers. While never an issue, it is an important point that Title X Command Authority actually skipped levels of leadership.

My personal view is that this puts an unnecessary strain on the Commandant and the young Company Tactical Officers. The good news, though, was that we had a great team which allowed for the sharing of information, mentorship, and the knowledge of when to create some space to allow for the Tactical Officers to execute their Title X responsibilities.

I am being unintentionally vague here as it is a hard thing to explain unless you experience it. I was not aware of any of this until I became the BTO, so there was a little bit of discovery learning.

Fortunately for me, I had a good onboarding experience. Even though I was coming from multiple leadership positions, and I had served at West Point previously, I did not know this side of West Point. I'm glad I was smart enough to recognize that! And the good news is that I had MAJ Sean B. and SFC Ray C. to teach me the ropes. I may not be the sharpest knife in the drawer, but I was smart enough to listen to these two guys.

The truth was that I thought I landed on the dark side of the moon when I got to West Point. There was some fallout from a recent "fist photo" in the *New York Times* and then the infamous pillow fight that occurred the previous year where some cadets got hurt. I once heard a saying that West Point is where 18-year-olds hide their laundry and 30-year-olds are paid to look for it. Obviously, it's a joke, but contrast some of this silliness — or at least what I perceived to be silliness — with the seriousness of the job I just came from at CENTCOM, where the stakes were literally life and death. I struggled a bit.

Sean and Ray helped with the transition. They helped me with perspective. They helped me prioritize where to put my efforts and what to push off for another time. They told me what would get me fired, and what would not. Most importantly, they reminded me that if I'm having a bad day, go and find a cadet and everything will be OK. And they were right.

The young people at the Academy were truly extraordinary in many ways. They definitely were not perfect, but they volunteered to be there and to serve their country. That counted for something.

The opportunities for mentorship and leader development are endless. While the cadets get plenty of focus and developmental opportunities — obviously, as the entire Academy is centered around them — I tried to spend more time with my Tactical Officers and Tactical NCOs. Together, they are called TAC Teams, or TACs.

With the rest of the rotating faculty at West Point, this entire group was referred to as the "second graduating class." It is this group of leaders who are rotating back into the force at the end of their assignments to be the senior NCOs, field grades, and battalion commanders managing and leading our Army. I felt that it was my duty to invest in them.

These TACs worked hard... really hard. I would argue that they had the toughest job at West Point. As a Brigade Tactical Department, we tried to maintain perspective — "no IED has ever gone off on 293" — and we did not want to be some sort of shadow government. In other words, let the cadets plan and execute their ideas and avoid pulling the override on their ideas. It's OK to let them run with scissors, as long as they don't fall and poke their eyes out!

In other words, short of a catastrophic meltdown or loss of life, limb, or eyesight, let them learn and grow from mistakes. The next step out of West Point is leading Soldiers, and we'd rather let them learn here at USMA where the stakes may not be as high. And as a TAC Team, while we weren't always successful and sometimes, we might have lost this perspective, at the end of the day we wanted to apply leadership to a problem as opposed to throwing a policy or regulation at the issue.

The more we focused on using all the tools in our leader

kitbag instead of quoting some sort of regulation violation, we believed we were teaching the cadets good habits and developing them as leaders.

Tragically, also like a Brigade, we experienced loss. Within my first 30 days as the BTO, two cadets died. Cadet Mitchell Winey, a member of the Class of 2018, was killed in a training accident on 2 June 2016 at Ft. Hood, Texas while participating in Cadet Troop Leader Training. He was one of nine Soldiers killed in this training accident.

Cadet Thomas Surdyke, a member of the Class of 2019, died on 28 June 2016. While on leave, Tom and a civilian were pulled out to sea by a riptide. Tom saved his civilian companion, but he later died. The citation from his Soldier's Medal reads: "Without regard for his own safety, CDT Surdyke immediately grabbed the civilian and physically assisted in keeping the civilian's head above water until help could arrive, before becoming overcome by exhaustion. Cadet Surdyke managed to push the civilian up, enabling a bystander on a paddle board to pull him out of the water, thereby saving the civilian's life."

I experienced plenty of death and loss in my life, but I had a hard time getting my head around the fact that this happened at West Point. It wasn't supposed to happen here. But it did. And while I did not know Mitch or Tom personally, I met their families, their friends, and worked with our team to make sure notifications were done correctly, memorials were professional and appropriate, and that we found ways to continue to honor their legacy. I was proud of our team and proud of our Cadets, and I was reminded once again that I was part of something bigger than myself.

I loved the TAC Teams — the officers and NCOs. They worked incredibly hard. They were at the tip of the spear. I greatly appreciated their humor, candor, and willingness to share with me their ideas and perspectives. Never satisfied with the status quo, they always looked for ways to improve their fighting position or more specifically, how to better develop cadets. The stakes were high. All eyes were on them — those that mattered and those that did not.

Short of combat, this was arguably one of the higher stress jobs in the Army. And every year we were rolling out another 1,000 Second Lieutenants (2LTs) into the force. We needed to get this right, or at least do the very best we could.

PUSMA AND AFGHANISTAN

I was a little bit of a unicorn in the Army. I was an operationally serving Infantryman with a Ph.D. By design — in terms of professional education and branch experience — this was an anomaly. That said, this did posture me for a professional opportunity. A little more than a year out from MRD, I had the opportunity to compete for a permanent position at West Point as a Professor, United States Military Academy, or PUSMA. This was for the position of Deputy Head, Department of Behavioral Sciences and Leadership. It was a tremendous opportunity; one only afforded because I had my Ph.D. This would allow me to stay on active duty, should I choose to do so, until age 64. After much thought and discussion with my family, I decided to compete.

This was by no means a guarantee. For one, I was already a little bit older than the other candidates — I was within 18 months of my MRD — and for another, I had not served as an Academy Professor at USMA (essentially, tenured faculty). Neither one of these factors was a negative against me, but it was a bit outside the norm.

In fact, when I eventually was selected, I was the only PUSMA who had never served as an Academy Professor. Overall, the selection process was a good one. I eventually advanced to the final series of interviews, and I ultimately was selected by the committee and then voted in by the Academic Board. The final step was Senate confirmation, which allowed me to be advanced to Brigadier General when I retired. Once this was complete, I was officially a PUSMA. All of this became official within a week of my MRD, providing some unnecessary stress in my life!

An interesting aside, this selection also came with a branch change. My "branch" was officially PUSMA, and I no longer wore the crossed rifles of an Infantryman. Several people asked me how that felt. From my perspective, it was no big deal. The younger me would have probably revolted, but with years comes wisdom... or so they say!

It's not the branch insignia, the blue chord, or what I wear on my uniform that made me an Infantryman. It's your spirit, what is in your heart. The soul of an Infantryman isn't defined by anything tangible. You either are, or you are not. Of course I was still an Infantryman. And now, because my MRD was pushed out to age 64, I could continue to serve.

One concern I did have was how relevant I was as an Army officer. Academically, I was fine. I was still publishing, reading, and teaching, and I was recently promoted to Associate Professor. Militarily, however, while engaged in training and staying current on doctrine and professional knowledge, I felt a bit dated.

I needed to do something to get back in the game operationally. My view was and remains that senior military faculty at West Point need to keep their professional acumen sharp so that they can continue to coach, teach, and mentor the rotating faculty — the Captains, Majors, and Lieutenant Colonels who are rotating back to the force.

I found that opportunity in Afghanistan with the SO-JTF-A/NSOCC-A and working on the personal staff of then-Major General Chris Donahue (Cd) — which takes us full circle back to the *Preface* of this book. I had known Cd for a while, so it was exciting to be a part of his team. An incredible leader, Soldier, and friend, I spent a little over three months in Afghanistan in the summer of 2019. As somewhat of an advisor to the Commanding General, I traveled with him, I was part of the most important conversations we were having about the current and future state of Afghanistan, and I worked in an organization — special operations forces — where I had not really been before. Of course, multiple times in my career, I worked with my SOF teammates.

This, however, was the first time I had the opportunity to be a part of it, day-in and day-out. I brought back to West Point a bunch of great lessons on leadership, operational and strategic decision making, and the tactical fight. Because of this experience, I was better personally and professionally.

DEPUTY DEPARTMENT HEAD AND ARMY FOOTBALL

Someone once said to me, "No one strives to be a deputy." For me, that was very true. I wanted to be the Department Head someday. It is a unique dynamic at West Point in that when the Department Head selects a Deputy, he/she is choosing one's successor. All that said, I was in no rush, and I was very happy to continue to teach and mentor cadets and mentor the junior/rotating military faculty.

In fact, it was with this latter group where I think I had the biggest impact. They were a hard-working group of young officers who were excited to be teaching cadets. Their enthusiasm was contagious. It has been said that "the lifeblood of an academic department is the rotating faculty." For anyone who spends a little bit of time in one of the academic departments at West Point, no truer words have been spoken.

I taught a broad range of courses from the core leadership course that all cadets take, to sociology courses, to management courses. My academic degree and background lent itself to this. But by far, the course I loved the most, and that was arguably the most impactful for the cadets, was *PL471: Leadership in Combat.*

In my eyes, this was the capstone leadership course. It was mainly only firstie/seniors in the class, and I taught it second semester, so it was very real for them. I loved the flexibility of adjusting the curriculum and course guide to meet the demands and the needs out in the force, while maintaining the academic rigor of a college-level course. I leveraged a broad range of speakers, books, research, and films to add additional content to the course.

Personally, the other thing that was very special about this was that I took the same course when I was a firstie at West Point. The course was taught by the department head, then-COL Howard Prince. A Vietnam veteran who was seriously wounded at the Battle of Hue, he had us read books like *The Centurions* by Jean Larteguy and *The Forgotten Soldier* by Guy Sajer. These were classics about Soldiers in combat and the human spirit, books I still have on my bookshelf.

While I was teaching my version of the course, I had the opportunity to reconnect with the now-Brigadier General (Ret) Prince. It was very special to be part of his legacy. I am glad I was able to tell him this. Howard Prince died on 19 May 2021.

Shortly after being confirmed and installed as a PUSMA, the Superintendent (SUPT) talked with me about being the Head Officer Representative (OR) for Army Football. There's a bit of a backstory to all of this in that I was going to be the Head OR for Army Hockey. The SUPT wanted me to be the

OR for football. While it may have been phrased as a question, in my Army world, when a three-star General either suggests or requests something, I'm moving to the sound of the guns. OK, maybe that's a bit dramatic, but the truth is, I became the Head OR for Army Football in June 2019, and I was excited about it.

Each intercollegiate athletic team at the Academy has an OR team. The size of the OR team is dependent on the size of the athletic team. Given that Army football is the biggest team in terms of the number of players and staff, revenue generation, exposure, and any other metric you wanted to throw at this, we had the biggest OR team. I can't recall the exact number, but I had somewhere around 15-20 ORs. These were officers, NCOs, civilian faculty, or USMA staff members. All were volunteers.

In general, our role was to provide mentorship, advice, and counsel to the cadet-athletes and to serve as somewhat of a consigliere to the coaches on a cadet's military, leadership, and academic development. I organized the OR teams by position. So, as an example, there was an OR for the offense and then an OR for the Offensive Line, Quarterbacks, Running Backs, and so on — same for the defense and special teams. I also provided OR coverage for the cadet-managers, cadet-videographers, and other administrative positions.

I've always had a difficult time explaining the intricacies of the duties and responsibilities of ORs. It's unique for sure. It should be a full-time job, but we were all volunteers. As the Head OR, the Head Coach was my counterpart. While I had relationships with all the coaches and football staff, it was Coach Jeff Monken who I talked to daily. I went to almost every practice. I was at all games — home and away. The only way I could properly do my job was to be "in the job." Similar to the mantra "The war isn't won on the FOB," I couldn't be a good OR sitting in my office in Thayer Hall. And all of this was in addition to my regular duties as the Deputy Department Head and teaching classes.

The cadet-athletes were hard working young men. In no way were they perfect, but to balance the demands of being a cadet and a Division 1 athlete was nothing short of extraordinary. And it wasn't just "being a cadet." It was having to excel in all three pillars of cadet development — academic, physical, and military. The nonsense of "corps squad get-over" is just that... nonsense. I'm proud of each and every one of these men who were like family to me.

Working with Coach Monken was a tremendous honor. Aside from being a great football coach, he is one of the best leader developers at the Academy. I greatly appreciated his candor, forthrightness, and how he sought my counsel on all matters related to cadet development. I believe we made a good team.

Name, Image, and Likeness — or NIL — doesn't exist at the Service Academies. The transfer portal only works one way, and that's out. So how do you build a winner at the highest level of college football? You build it by recruiting the right young man who represents the Academy values. Young men who are coachable; young men who are tough; young men who are winners; young men of character. And then you coach them. You coach them to do their job. You coach them to master the basics. You coach them to never take a play off. You coach them to be the toughest and fittest team on the field. West Point, and by extension Army football, teaches you that good leaders master the basics and do their job.

It was about this time that COVID changed the world as we knew it. Like every other Army unit and college/university, USMA navigated this space as best we could to take care of our people and accomplish our mission. I believe we emerged on the back end of this stronger and better as an institution.

My COVID story began on 12 March 2020. It was a Thursday. The cadets were on spring break, due back in a few days. There was a lot of uncertainty about the way ahead. My solution to the problem was to go climbing in the Adirondacks. The plan was to meet up with my friend and New York City firefighter, Matt M., at a hostel in the Adirondacks on Wednesday evening and then climb on Thursday and Friday and head back. And that's what we did... sort of.

Matt and I crashed at the hostel on Wednesday night and got up early Thursday morning and grabbed breakfast at a local diner. We took his truck to the diner and then trailhead. I left my truck at the hostel since we planned on coming back that evening. Again, COVID was happening, whatever that meant, but it really didn't impact anything we were planning.

After breakfast, it was a short drive to the trailhead, and we began our hike and then ascended the North Face of Gothics. The North Face of Gothics is one of the most conspicuous landmarks of the Adirondacks High Peaks, drawing the eye if you're in downtown Lake Placid, New York. It's a wall of rock and ice (in the winter) roughly 1,200 feet high and a quarter-mile wide. Yes, the North Face is big, and if you want to climb it, plan on a big day. We had planned on a big day... we just didn't know how big it was going to be!

About three-quarters of the way up, Matt had me on belay. He was maybe 50 feet above me. It was not a super technical climb, but I was somewhat of a novice, and I was taking it nice and easy. The combination of my ice tools and crampons helped me navigate the ice. At one point, however, I lost my footing. The good news was that Matt had me on belay, so I wasn't going to slide down the mountain. The bad news was that I swung off to the right. My line of drift as I was climbing took me to the left of where Matt was positioned above me. I should have paid more attention to this. When I lost my footing and swung back to the right, which would have put me directly below Matt, my crampon on my left boot caught an exposed rock and snapped my ankle. I screamed like a bitch.

Fortunately, I couldn't have been with a more experienced person. Not only was Matt a very experienced firefighter, he was also a very experienced climber and a certified mountain guide.

Without going through the play-by-play of what happened next, Matt and I ended up downclimbing until we could get into cell phone service. Mind you, it's March in the Adirondacks so the temperature was dropping fast, and it was getting dark. The Park Service sent a Vietnam-era looking helicopter out to us. There was nowhere to land given the angle of the terrain, so a Ranger rappelled down to our location, assessed my injury, and made the call to hoist me out and get me to the hospital in Lake Placid. And that's what happened.

It had been a long time since I was suspended underneath a helicopter (the first time was in 1988 at Jungle School in Panama!), but it was like riding a bike! I was hooked up and hoisted into the helicopter and then flown to the Emergency Room at Lake Placid. Matt and the Ranger had to hike out back to the trailhead. Not an easy task!

At the hospital, the doctors assessed my injury and wanted to put me in surgery immediately. I declined. I needed to have this done back at West Point. So, the doctors put a cast on my foot up to the knee and wished me good luck. Matt gave me a ride back to the hostel to get my truck. Fortunately, it was my left leg so I could drive. With no pain medication, very little to eat or drink, and roughly six hours ahead of me, I began my drive. I was a safety officer's worst nightmare. Matt followed me in his truck almost the entire way. I got home at roughly three or four in the morning.

Later that morning, my wife found me laying on the sofa in the living room, still in my climbing clothes. This was Friday, 13 March 2020. It was the day that I remember the world shut down because of COVID. A few weeks later, I had surgery. I had the entire hospital to myself. I now have a plate and eight screws holding my foot together. The orthopedic surgeons did a great job.

Matt M. saved my life that day. It sounds a little dramatic, but there is no way I was getting out of there without him. What stands out to me was how calm he was, professional, and deliberate in his decision making. We had been out together before, so I trusted him. I literally trusted him with my life. There's nothing flashy or outlandish about what he did that day. He quite simply was a professional. He was a Quiet Professional and to this day the qualities he displayed on the North Face are ones I try to emulate in my own life and in tough situations.

USMA CHIEF OF STAFF

Never in a million years did I ever think that my last job in the Army would be as the USMA Chief of Staff. Certainly, as a cadet, I had no idea that this was even a thing. But in January 2022, this is where I found myself... in the Chief's office.

I think it's easy to forget that the United States Military Academy is a three-star Army Command with a three-star Headquarters with three-star responsibilities. Unlike other three-star commands — XVIII Airborne Corps, III Armored Corps, U.S. Army Installation Management Command — there is generally a four-star Command above them as a Higher Headquarters (HHQ). And the staff of these three-star headquarters is resourced to accomplish specified and implied missions and tasks. It's a little different for USMA.

For one, our HHQ is the Department of the Army. The

Secretary of the Army and the Chief of Staff of the Army are the Superintendent's bosses. In parallel, the USMA staff counterparts are the Army staff. My direct counterpart was the Director of the Army Staff, or the DAS. He was a three-star General Officer. I was a Colonel. My point here is that our reporting requirements were driven directly by Big Army. There was no four-star Command and Staff—i.e., Forces Command (FORSCOM), Training and Doctrine Command (TRADOC), etc.—to take on any of this for us. And all our resourcing requests went directly to the Army staff. When we needed to accelerate the order for Cadet Gray Jackets, this went directly to the Army G4!

Additionally, we had other stakeholders who were invested in the Academy and very much part of the governance structure—either stated or implied. This included most importantly the Board of Visitors, which consisted of members of Congress and Presidential appointees. And then there was the Long Gray Line, the alumni. While not directly part of any governance structure, they were an influential group invested in the Academy.

To manage all of this was a USMA Staff that punched way above its weight class. Our Army designated organizational structure didn't allow for much in terms of our authorized personnel strength, but we did build in some much-needed staff positions by taking from elsewhere at the Academy. There was a risk to this reprioritization of slots—"robbing Peter to pay Paul"—so we needed to be judicious in how we approached this. We also needed to be selective in our talent acquisition. We got it right most of the time. These were civilians, officers, and NCOs who were innovators, decision makers, and problem solvers. As an example, our G8 shop, led by Melissa C., was the best I've ever seen in the Army. Although under-resourced from a personnel perspective, they managed a complex budget with a small team of talented professionals.

As the Chief, my job was made easy by having a first-rate staff. The challenge we faced was managing the myriad of tasks related to running an installation; supporting multiple subordinate commands (to include the Garrison at Ft. Hamilton, NY); resourcing the Academy (which is different from running an installation); working with the Army staff and across other Army organizations and external stakeholders; and serving as the Superintendent's staff, which involved the processes and systems of all the administrative support that goes into this.

There is no doubt I am missing something. The complexity of running and operating the United States Military Academy is not something I'm sure I can do justice to. All that said, it was incredibly rewarding, made ever more so by the quality of people with whom I served.

One of those people was the USMA Command Sergeant Major, Mike C. Mike and I had a history. When I was the 1-24 IN Battalion Commander, Mike was my Scout Platoon Leader. We initially did not have an officer in that position, so we gave the job to Mike, then a Sergeant First Class. His job was to rebuild the Scout Platoon after the 16-month deployment to Iraq and the multiple departures due to Soldier rotation. A few months into command, Mike was selected for E8, Master Sergeant, and was promoted to First Sergeant. He ended up moving to another Battalion. All that said, he was a great NCO and became a very good friend.

As the Chief of Staff, there was no better teammate as the CSM than Mike. Our offices were right next to each other and there was a door that allowed us to pass back and forth undetected by others! It was Mike's wise counsel, calm and deliberate decision making, and his unwavering professionalism that helped me (us) navigate many difficult situations.

"Steady in the Saddle" was a phrase he used often to help keep the team focused and on an even keel. Just as a horse gets rattled when its rider is rattled, so it is for others when leaders are rattled. We were all better for the time we spent with Mike, and I couldn't ask for a better Ranger-buddy when it came time to retire. My time as the Chief of Staff came to an end in May 2023.

8. LAST WORD

Years ago, a mentor once told me that you'll know when it's time to retire. That's exactly what happened to me. I could have stayed on active duty for another seven or eight years. I just wasn't sure I wanted to hang around that long. There was more to it, but it wasn't all that complicated. It was just time.

I retired from the Army on 1 September 2023. For all the math majors out there, that's not quite 35 years. It's 34 and some change. However, I'm giving myself a little extra credit. I was on active duty from 1985 to 2023, but those four years at West Point (1985-1989) don't count toward retirement. So, I guess it's closer to 38 years of active-duty service. All that to be said, when I was commissioned in 1989, I could never have imagined serving as long as I did. My career was unremarkable, but I did serve with remarkable people.

I was a Soldier for a long time. And in that time, I determined that there are no guarantees in life and that things are not always fair. But the one truism that has defined my view on being a Soldier is that there is nothing better than being part of a great team, with people you care about; people you love and respect, and who make you better each and every day. Iron sharpens iron.

And greatness isn't what just happens when you show up

for formation, drive through the gate, or deploy to a distant battlefield. Be a great husband, wife, father, mother, son, or daughter and I promise, you will be an even better Soldier.

As I close this out, I have one final thought. To quote from the book, *Legacy*, the story of the world's most successful rugby team, the legendary All Blacks of New Zealand:

> "There are no crowds lining the extra mile. On the extra mile, we are on our own: just us and the road, just us and the blank sheet of paper, just us and the challenge we've set ourselves. It's the work we do behind closed doors that makes the difference out on the field of play, in whichever field we compete, whether we're on a team, leading a business, or just leading our life."

So, my final message as a Soldier is this. There are no crowds lining the extra mile. Champions do extra... in all aspects of life, not just what matters to them most. Be a champion.

DUTY — HONOR — COUNTRY

I'm on the left. D4 Sandhurst Team 1987.

The Legion TAC in Iraq 2008-2009. Great Americans.
I'm standing to the far right.

I'm on the right. Mick is in the center. Winter PT in Alaska!

*Pinning the Bronze Star with Valor Device
on Stephen Stoops in 2012.*

*Raider 6 in the center. I'm to his left.
Night patrol in Iraq 2003-2004.*

With the legend, COL (R) Lew Millet.
Schofield Barracks circa 1995.

With SFC W. circa 2018. Legion 7 and I made him a Squad Leader in 2006 when he was a PFC.

2LT Reed with Santa at Panzer Kaserne
in the B/1-16 Company Area 1990.

I'm to the left, Raider 3. With Raider 6, GEN Odierno, and Raider 7. Late night 13 December 2003 at the Ironhorse Division Headquarters with the $750,000 that Saddam Hussein had with him the night he was captured.

ABOUT THE AUTHOR

Brian Reed, Ph.D., is a retired U.S. Army Brigadier General with over three decades of service as an Infantry officer. A graduate of the United States Military Academy, he commanded at every level from platoon to a Stryker Brigade Combat Team, including deployments for Operations Iraqi Freedom and Enduring Freedom. In addition to his Ph.D. in Sociology, Brian served on the faculty at West Point and later as the Academy's Chief of Staff. He concluded his military career as a Senior Advisor to the Secretary of Defense. and is now the founder of Wolf's Edge Consulting (https://wolf-sedgeconsulting.net/) where he specializes in leadership for high-stress environments.